Praise for Russell Shorto's

DESCARTES' BONES

"Shorto leaps from one intriguing topic to another, doing it with verve. . . . [He] delves into the oddest and most unappetizing aspects of Descartes' story along with the inspiringly lofty ones. . . . His insights are keen. And he is as drawn to great, overarching ideas as he is to historical factoids. Descartes' posthumous journey happens to be rich in both."

—*The New York Times*

"Hard to put down. . . . Few journalists have both the nimbleness of mind and the lucidity of prose required to explain abstract ideas; fewer still can tell a really good story. One sterling exception is Russell Shorto."
—*The New York Observer*

"Shorto offers the bizarre tale of the titular bones' grave-hopping over the decades as a metaphor for humanity's journey from medieval thinking—grounded in the authority of king and church—to modernity's emphasis on individual reason as the basis for morality and politics. You may recall [Descartes] from philosophy class . . . but it's a safe bet that you didn't have his ideas or enemies this clearly or exuberantly explained."

—*The Boston Globe*

"With the fascinating *Descartes' Bones*, Russell Shorto has produced another compelling intellectual detective story, one that illuminates the present as much as the dusty past."

—Jeffrey Toobin, author of *The Nine*

"Shorto uses the weird and convoluted story of what happened to the great philosopher's remains as a route to exploring Descartes' philosophy and influence. . . . [He is] in full command of the complexities of his story."   —*South Florida Sun-Sentinel*

"This is a beguiling book about the architecture of the way we live now. As Russell Shorto points out, Descartes is claimed by both the ferociously secular and the ferociously religious, but the truth is more complicated. The sooner we recognize that the world is too wild to be reduced to glib categorization, Shorto writes, the sooner we may be able to find ways to talk to, rather than yell at, one another."   —Jon Meacham, author of *Franklin and Winston* and *American Gospel*

"A pleasure to read. . . . Shorto relates the life of Descartes and the bizarre story of his remains with infectious relish and stylistic grace."   —*Publishers Weekly*

"A fascinating, colorful, and very readable account of early modern ideas and personalities. Shorto has a gift for storytelling. He brings the seventeenth century to life while doing justice to the philosophy."   —Professor Steven Nadler, author of *Rembrandt's Jews* and *Spinoza: A Life*

RUSSELL SHORTO

DESCARTES' BONES

Russell Shorto is the bestselling author of *The Island at the Center of the World* and a contributing writer at *The New York Times Magazine*. He lives in Amsterdam.

www.russellshorto.com

ALSO BY RUSSELL SHORTO

*The Island at
the Center of the World*

Gospel Truth

Saints and Madmen

DESCARTES' BONES

DESCARTES' BONES

A
SKELETAL
HISTORY
of the
CONFLICT
Between
FAITH AND REASON

RUSSELL SHORTO

VINTAGE BOOKS
A DIVISION OF RANDOM HOUSE, INC.
NEW YORK

FIRST VINTAGE BOOKS EDITION, SEPTEMBER 2009

The Library of Congress has cataloged the Doubleday edition as follows:
Shorto, Russell.
Descartes' bones : a skeletal history of the conflict between
faith and reason / Russell Shorto.
p.   cm.
Includes bibliographical references and index.
1. Descartes, René, 1596–1650.   2. Faith.   3. Reason.   I. Title.
B1875.S495 2008
194—dc22
2008001932

Vintage ISBN: 978-0-307-27566-0

Author photograph © Jennifer May
Book design by Lovedog Studio

www.vintagebooks.com

Printed in the United States of America
10 9 8 7 6 5 4 3 2 1

For my mother

"what can we bequeath
save our deposèd bodies to the ground?"

—Richard *II*, III , 2

Contents

Preface

PHILIPPE MENNECIER, THE DIRECTOR OF CONSERVATION AT the Musée de l'homme, the great anthropology museum in Paris, is a tall, narrow man, thin of hair, with wire-rimmed glasses and the aspect of a bird of prey. Suitably enough, his office is something of an aerie, a low-ceilinged, rectangular box built as an afterthought on the roof of the museum's headquarters, which you get to by climbing a portable metal staircase. Up here, he has surely one of the grandest workplace vistas on earth, taking in much of the Paris skyline. The view also gives a metaphorical frame to the work that Mennecier and his staff do: on one side, so close you can't get a full picture of it, is the Eiffel Tower, that nineteenth-century obelisk to reason and order; on the other is the Passy Cemetery, one of those wondrous Parisian cemeteries that, with their tangle of paths and tombs and high surrounding wall, resemble medieval cities in miniature, but ones populated by the dead rather than the living.

Death and order: that sums up the work that goes on here. The museum is not on the standard tourist itinerary, but it's a place for which the French have a particular fondness. It was

founded in the early nineteenth century, as part of the first burst
of enthusiasm for the search for human origins, when explorer-
scientists—hale, moustachioed, fanatically dedicated—combed
the far reaches of the earth for anthropological specimens and
human remains. Reflecting those origins, the museum has a retro
feel. You might think of it as a temple devoted to the cult of evolution, which brings reason to bear on the conundrum of life and
death by using bones to tell the modern story of who we are and
how we got here. The cemetery below, meanwhile, with its mute
stone crosses, gives another version.

Echoing the views and their bookended representations of reason and mortality, Mennecier's office is cluttered with computer
equipment and human remains—a tray randomly placed on a
shelf neatly holds six human skulls, as if six were a standard set.
But Mennecier is not himself an anthropologist but a linguist,
as he made a point of noting when we first met. And what languages are his specialities? *"Esquimau et russe,"* he declaimed with
a flourish: Eskimo and Russian. To properly appreciate this response you should know that it had already been established that
he didn't speak English. What could be more exquisitely right for
a French linguist than that he profess no working knowledge of
the world's dominant language but be one of the leading experts
on the Eskimo dialect spoken exclusively in eastern Greenland
and author of the only Tunumiisut-French grammar? On top of
which, his pursuit of Inuit language variations around the earth's
northern reaches eventually led him to Siberia, so that he became
fluent in Russian and now, as a sideline, translates contemporary
Russian novels into French.

All of which is to say that Mennecier is a French intellectual.
To many people in this age of universal dumbing down, that
would be considered a slur, suggesting things like arrogance and

a focus on narrow, cerebral, navel-gazing concerns. But the term can also encompass a way of looking at the world that is becoming sadly rare—call it a serious commitment to idiosyncrasy. People who are configured this way can give you a headache, but they can also delight you with their inexorable weirdness. They work the way a joke does, pulling you unexpectedly off the easy chair that is your customary vantage point. They make you remember, if only for a moment, what a wild place the world really is. So I was happy to ride this wave for the next few minutes, to listen to a little discourse on the seven Eskimo dialects, how they divide into two families, what linguistic markers separate them, and the efforts to preserve the dialects and their cultures.

Eventually, we clattered back down the metal staircase to the floor below. Here, two women in lab coats sat at a table handling human bones: long leg bones with porous, knobby joints; skulls of a slightly sickening orangey-brown hue. In the next room we passed a group of maybe four dozen complete human skeletons hanging on hooks, with a single gorilla skeleton standing in front of them like a short, thick sergeant drilling a lanky squadron. As we went back out through the entry to this area, we walked by a bust of Pierre-Paul Broca, the nineteenth-century anthropologist and pioneer of brain study. We headed downstairs, passing the main exhibition floor of the museum, with its quaint permanent exhibition, an almost aggressively confident display devoted to human evolution, in which a succession of dramatically lit dioramas hit the milestones in the march of bipedalism, from *Australopithecus*, with its wide arching plates of bone across the eyes, to Cro-Magnon, with its voluminous cranial capacity and frontal bulge, to their more delicate modern cousins.

Finally we could descend no further. The basement was in the process of being remodeled, and the fresh plasterwork and exposed bulbs gave it the pleasingly appropriate feel of a catacomb. My host produced a set of keys and opened a storage room door. Once we were inside, he unlocked a wall cabinet and pulled out a finely polished, curiously elegant wooden box whose lid was fastened with metal hasps. He unsnapped these; there was a flourish of gauzy white paper, then he reached in and extracted the object I had come to see.

It was small, smooth, surprisingly light. The color varied: in places it had been rubbed to a pearly gloss while other areas were deeply grimed; but mostly it had the look of old parchment. And indeed, it was an object with stories to tell, not only figuratively but literally. More than two centuries ago someone had written a lofty poem in Latin on its crown, the letters of which were now faded to a watery brown. Another inscription, right across the forehead, hinted darkly—and in Swedish—of a theft. Tightly scrawled signatures of three of the men who had owned it through the ages were faintly visible on the sides. It was the skull of René Descartes, the so-called father of modern philosophy and one of the more consequential humans who ever lived. Mennecier set it on a table before me. *"Voilà le philosophe,"* he said drily.

T HREE YEARS EARLIER, while sitting in the Main Reading Room of the New York Public Library plowing through a volume of seventeenth-century philosophy, I had chanced upon the fact that, sixteen years after his death in 1650, Descartes had suffered the indignity of having his bones dug up, after which people began taking pieces of his remains.

Why do certain thoughts stick in the mind? They seem to have no practical value but stand out from sheer strangeness. Typically you'll entertain them for a while, like a child's toy found between sofa cushions, then forget them, uselessness having defeated novelty. Certainly this item about Descartes' bones seemed a pristine example of a useless piece of information. And yet I fell in love with it, the way you can only fall in love with something truly odd that you find buried in a very old book. It has happened to me only a few times: you have the feeling, improbable but strong, of having uncovered a dormant seed, one that was planted in just that spot by someone now long dead who knew, or hoped, that one day you would find it, water it, and bring it to life.

So I pursued it, first in off moments, in books, then, as it took hold, moving my family to Europe for a year, where I spent long days in the postmodern cloisters of the Bibliothèque nationale in Paris, contacted philosophers and historians, traveled from the house in the Loire Valley where Descartes was born (which still stands) to the house in Stockholm where he died (which also still stands), and followed the trail of the bones across western Europe. Eventually I found myself standing in the basement of a museum in Paris, staring into a skull's blasted eye sockets, like Hamlet contemplating poor Yorick.

As I investigated, the story of Descartes' bones unfurled before me and stretched across the centuries, and it revealed itself to be more than a curiosity. Today Descartes is most readily thought of as a mathematician—the inventor of analytical geometry—and as creator of the modern philosophical puzzle of dualism, which holds that the mind and its thoughts exist in a different category or somehow on a different plane from the physical world, so that neither can be translated into the other or understood in terms of the other. On this score, he has long since been put in his place: the prevailing wisdom in neuroscience and philosophy is that Descartes was dead wrong in conjuring up his two dissimilar substances. Mind and body—mind and brain—aren't fundamentally different after all. This notion has all sorts of consequences, which are being explored by philosophers, linguists, spiritual thinkers, computer scientists, and others.

But during his lifetime, and in the decades following, Descartes loomed larger. He was seen by many of his contemporaries as the man who laid the intellectual foundation for the whole modern program, which grounds everything from morality to law to politics and social organization on reason and the individual perception of reason. There is truth in this view of Descartes' influence. His famous "method"—which involved questioning assumptions, taking no assertion on faith, and building our understanding of the world on provable observations rather than tradition—became the basis for the scientific method. His reorientation of knowledge so that it was no longer based on collective authority (what the king decrees, what the church demands) but on a newly empowered *self*—the individual mind and its "good sense"—became a starting point for the development of democracy, psychology, and much else that we think of as modern.

What I began to realize was that people who lived in the generations that followed Descartes treated his bones as symbols—relics—of the new turn the world had taken. Yet, because they differed as to what this new turn was and what it meant, they treated the bones in different ways. The story on which I became fixated—small, weird, serpentine, insignificant—intersects some of the grandest events imaginable: the birth of science, the rise of democracy, the philosophical mind-body problem, the ongoing confusion over the terrains of science and religion. The story crisscrosses Europe and encompasses people from all walks of life—Louis XIV, a Swedish casino operator, poets and priests, philosophers and physicists—as these people used the bones, stole them, sold them, revered them, tussled over them, passed them from hand to hand.

Yet it wasn't until two or more years after my first exposure to the fact that Descartes' bones were dug up and passed around that I got an inkling of the real source of my interest. In college I studied Western philosophy. Like numberless humanities undergraduates before and since, I spent those four years reveling in the work of philosophers, poets, novelists, artists: the men and women who created the mental space that I have lived in ever since, the architects of the modern mind.

Many of us used to think that "modern" was a given, a common ground. And by modern I don't just mean the big things—science, reason, democracy—that we associate with the word but all the reactions to and offshoots of these concepts, too, from Romantic poetry to the Sex Pistols, from Internet dating to hedge fund trading. For better or worse, all of this is somehow bound up together and tied to who we are—and mainly, we think, it's for the better. Don't we?

Apparently a lot of people do not. Today, the very idea of modern society—which, at least in theory, relies on the tool of reason and notions such as equality to solve problems and lumber forward—seems to be under assault from several directions. Islamic terrorism, which is not just anti-Western but antimodern, is of course a dominant concern in the West, but other forms of religious intolerance—Christian, Jewish, Hindu—seem to be flourishing as well.

If these constitute a right wing of attack on modernity, there are other threats. Within secular Western society there are those who say that modernity is passé, that in a postmodern world developments such as globalization, the Internet, and asymmetrical warfare mean that the old verities of the modern era—the idea of "progress," for example: the notion that you can get a reasonably objective view of things and then make decisions and move forward to something better—are out the window. To some, modernity has come to be synonymous with colonialism, the exploitation of non-Western peoples, the use of science and technology for inhuman purposes, environmental catastrophe. Many secularists also see religion itself as an enemy, arguing that it promotes war, division, and prejudice. Responding to the upsurge of faith-based fundamentalism, Richard Dawkins, Christopher Hitchens, and others have written secularist manifestos against religion, some of which have become best sellers.

In the perennial conflict between faith and reason, we tend to think of the one as old and the other as new, but today both the left and the right rely on Descartes. His remains—his metaphorical remains but also his actual bones—are so elemental that both of these competing camps put them to use. It's not surprising that the archetypal modern philosopher would be godfather to the left; since Cartesianism was based on doubt, on questioning everything until you reach a bedrock of fact, it can be seen as the root not only of the scientific method but of self-government, the modern idea of individual rights, and of the equally modern distrust of authority. At the same time, another element of Descartes' philosophy—what is known as Cartesian dualism, the notion that our minds (and souls) exist separately from the physical world—has been embraced by the right. Conservative thinkers—monarchs, theologians, philosophers—have followed Descartes' mind-body distinction to buttress their argument that there is an eternal realm of thought, belief, and ideals that can't be touched by the prying fingers of science and that human morality and earthly power are grounded in this timeless sphere.

Most people seem to be caught between these tidal currents—the pull to faith and tradition in a dangerous world, the argument that religion is at the core of the world's problems and only a revived commitment to individual freedoms and rights will steer humanity into a better future. They are troubled by religious fundamentalisms, with their dead and deadly certainty. But they can appreciate some of the criticisms of modernity, both from the right and from the left. You might say that there is not so much a split as a three-way divide in the world today. Colin Slee, the dean of the Anglican diocese of Southwark in London, put it this way in talking about the new society that he sees coming into being in England: "You have a triangle with fundamentalist secularists in one corner, fundamentalist faith people in another, and then the intelligent, thinking liberals of Anglicanism, Roman Catholicism, baptism, methodism, other faiths—and, indeed, thinking atheists—in the other corner."

If the West is heading toward some kind of crisis, it's worth asking ourselves a few basic questions. Modern society as we normally define it—a secular culture built around tolerance, reason, and democratic values—occupies a rather small portion of the world, and there are signs that it is shrinking. Is modernity the inexorable force of progress that we tend to assume? Is it a mere moment of human history that is fast fading? If it is something to value, how can we rediscover it, separate the good and the bad in it, make it relevant and vital?

Eventually it occurred to me that the trail of Descartes' bones was a path through the landscape of the modern centuries. Following the bones was a way of retracing my own intellectual upbringing, reminding myself of what we've been through in the past four hundred years. This book is not an exhaustive survey of modernity but a record of a journey. It proceeds from the conviction that idiosyncrasy is a serious business.

What's more, its focus on bones isn't accidental. Following the trail of Descartes' bones showed me that philosophy, which we think of as an arid discipline, isn't all abstraction but is braided into human history and comes not just out of the human mind but out of the human body as well. Abstract thinking is an excellent and necessary tool, but the loftiest thoughts are rooted in our physical beings: in the curious way our hearts that love are entangled with our hearts that pump blood, in the fact that we die. While it's not a biography, the story does involve a man—whom history has caricatured as virtually a disembodied human brain but who turns out to be surprisingly full-blooded and substantial. Indeed, there is a sense in which Descartes' philosophy, for all its abstraction, sprang from places of human warmth: from his own body, for one, and also from the love he felt for the person who mattered most to him. It was a small, tender love that, in its quiet intimacy, nearly—but not quite—escaped the prying lens of history. Maybe it's true of every human exploration: dig deep enough and you find a love story.

That said, it shouldn't be surprising that we begin not with love or history or philosophy but with death.

DESCARTES' BONES

The Man Who Died

ON THE SOUTHERN EDGE OF STOCKHOLM'S OLD Town stands a four-story building that was constructed during the busy, fussy period called the Baroque. Its redbrick façade is ornamented with sandstone cherubs and crests. Two upright cannons flank the entry; bearded busts gaze down sternly on those who approach the door. If you could somehow ignore the designer handbag shop and the upscale "Glenfiddich Warehouse" restaurant/bar occupying the ground floor, and the streams of tourists moving past on a summer afternoon, the structure would probably seem very much of the year—1630—when a merchant named Erik von der Linde built it.

In the dead of night in the dead of winter in the year 1650, the most solemn rite of passage was playing out on an upper floor of this building. People hurried between rooms, past windows that looked out onto the dark, icy harbor below, exchanging information and worried looks. But if the occasion was grave, it wasn't quiet. For someone close to death, the man who lay in bed—not quite fifty-four years old, small-boned, ashen, the center of everyone's attention—was alarmingly active. It was fury that gave him these last bursts of adrenaline. His friend and protégé Pierre Chanut, the French ambassador to Sweden, in whose house he lay dying, was at his side constantly, trying to manage the man's anger while feeling doubly guilty: it was he who had urged René Descartes to come to this frozen land and he who had first contracted a fever, through which Descartes had nursed him before catching it himself.

Chanut fervently believed that Descartes was in the process of transforming the world with his revolutionary thinking. In this he was essentially correct. A change took place in the middle of the 1600s. People began to employ a new, sweeping kind of doubt, to question some of their most basic beliefs. The change was in a way more profound than the American and French revolutions, the Industrial Revolution, or the information age, because it underlay all of them and affected the very structure of people's thought—the way they perceived the world, the universe, and themselves in it. And the person most closely identified with this transformation was the man who lay dying in the Swedish winter. Pierre Chanut couldn't have known the scope of the future, but he knew, as did many others, that something staggeringly significant was afoot and that Descartes was at its center. It had by now dawned on the diplomat that, in bringing the philosopher here, he had unwittingly engineered a catastrophe.

The fever had given way to pneumonia; the patient's breath was ragged, his eyes wandering. Chanut had wanted to call the court physician, but Descartes raged against that idea. Finally, from her fairy tale palace on the other side of the small island in the harbor that was the center of Stockholm, Christina, the twenty-three-year-old queen of Sweden, who would go down as one of the more remarkable personalities in European history (there is, for starters, the centuries-old line of serious speculation that she was in fact a king), sent her physician to attend him. It was Christina who, with Chanut, had coaxed the intellectual celebrity northward in the first place.

The doctor, a Dutchman named Wullens, approached the bed reluctantly. There was a sharp exchange in which the philosopher made it venomously clear he thought the physician an ass. The encounter climaxed when Wullens proposed bloodletting, whereupon the patient erupted with a theatrical cry—"Gentlemen, spare French blood!"—and ordered the man out. Wullens departed, washing his hands of the business, muttering as he went a rather fatuous piece of consolation from the Roman poet Horace: "He who saves someone against his will does the same as to kill him."

The rage had two components. First, the philosopher had known Wullens during his long years in the Dutch provinces. One of the early public airings of Descartes' philosophy had come at Leiden University, and it caused an uproar among those who considered it a challenge to the whole system of education and thought that had been in existence throughout Europe for centuries. Wullens had stood with those who opposed the new philosophy. Descartes never forgot an enemy.

But there was another reason for the anger. In a peculiar way, much of Descartes' career had been a kind of chess match with death, and for a long time he had actually convinced himself that he had the upper hand. He had been a sickly child, with a pale complexion and a dry cough that he had inherited from his mother, who died when he was a year old. His father—a jurist and a man of power and ambition—seems to have despised the child's weakness and favored his older brother. The family doctors didn't bother to hide from the boy their conviction that he would die young.

When he was ten, however, Descartes was sent off to the Jesuit college of La Flèche in Anjou, one of the finest educational establishments in Europe. There, to his surprise, he flourished. He became strong, healthy, vigorous, aware of the wider world, and hungry for knowledge. But the early experience remained lodged inside him. When he settled into his mature work, medicine became its central focus. He developed his revolutionary philosophy, with its grounding not in the Bible or ancient writers but in human reason, and became famous and infamous for it. But the heart of it, the deep reason for it, was his desire to solve the puzzle of the body, to cure disease, and to lengthen human life—including his own. At the end of the *Discourse on the Method*, his epoch-changing work of philosophy, he vowed to the reader not that in the future he would craft a revised metaphysics or a new approach to mathematics but that he would "devote what time I may still have to live to no other occupation than that of endeavoring to acquire some knowledge of Nature, which shall be of such a kind as to enable us therefrom to deduce rules in medicine of greater certainty than those in present use." Five years before he lay dying in Sweden he wrote to an English earl, "The preservation of health has *always* been the principal end of my studies."

The same goal was in the minds of many of his contemporaries. When we think of science and the spark of modernity, we tend to think of astronomy: Galileo crafting his telescopes and peering into the skies above central Italy; locating sunspots, moons around Jupiter, craters on the earth's moon, and other irregularities in a universe that the church had taught was perfect; amassing data that corroborated the theory that the earth revolves around the sun; encountering the systematic wrath of the Inquisition. In our perennial effort to understand who we are and what it means that we are "modern," we choose astronomy as a starting point in part because it provides a sturdy metaphorical peg for thinking of the massive change that humanity underwent in the seventeenth century, when we—seemingly—left our mythic, biblical selves behind and reoriented ourselves in the cosmos. In 1957—the year of Sputnik and the dawning of the space age, a time when people had a simpler, clearer sense of "modern" than they do today and felt ready to embrace what they thought the word meant—a bestselling book expressed this idea in its title: the change was "From the Closed World to the Infinite Universe."

But one could just as easily see modernity springing from the intense interest in the human body that arose in Europe at the same time. If our place in the universe is an elemental marker of who we think we are, our physical being is something more. The magnitude of human suffering down the centuries is somewhat quantifiable. The life expectancy of a child born in Descartes' France was twenty-eight; in England between 1540 and 1800 it was an estimated thirty-seven. Similar rates—in the twenties and thirties—held for high-born citizens of ancient Rome, forager societies in Africa and South America, and people in rural India and China into the early twentieth century. More than half of all children born in London around the time of the American Revolution could be expected to die by age fifteen. And most deaths in early modern Europe were caused not by war or marauding brigands but by disease. Century upon century, hour after desperate hour, parents watched helplessly as their children succumbed to maladies whose very names—ague, apoplexy, flux, dropsy, commotion, consumption—spoke of the misty ignorance that was a definitive sentence.

The mists have lifted somewhat in three and a half centuries—we live longer and healthier lives—and still the body remains a touchstone of modernity. Zoloft, Lipitor, Viagra, Botox,

ibuprofen, angioplasty, insulin, birth control pills, hormone replacement therapy, anabolic steroids—we don't merely allow science and technology into our physical beings but insist that they
continually do more to better manipulate and aid the brute facts
of our flesh and blood and bone selves. Embedded in this outlook
is an idea of the body as a machine, so that illness is seen as a
breakdown of the machine, healing involves repairing the broken
parts, and a doctor is a kind of mechanic with medications as his
or her tools. This simplistic view has been changing in the last
twenty or thirty years. We have a hankering now to see mind
and body as deeply connected, to appreciate the way thoughts and
the environment influence our physical being. Yet the mechanical
model has been very successful, and our medicine is still largely
constructed around it. And it was this model that came into being
in Descartes' generation.

This new way of viewing the human body was bewildering
when it was first aired. Many people, in fact, equated it with atheism. It was frankly at odds with the overall approach to knowledge
in the period against which modernity arose. Aristotelianism, or
Scholasticism, was a blend of Christian theology and thinking
derived from Aristotle and other ancient Greeks. These streams
of thought had stewed together for centuries and resulted in a
worldview that, often spiced with astrology and folklore, treated
every subject under the sun, from the story of creation to the roles
of men and women. It explained why a stone dropped from a window fell to the earth rather than floating upward (because objects
want to move toward the center of the earth, which is the center
of the universe); it told what happened when you died; it gave an
account of the end of all things.

The premodern medical establishment—which Descartes

had dedicated himself to overthrowing—was built around the teachings of the ancient Greek physician Galen, whose work in turn was dependent on Aristotle's division of the physical world into the four elements of earth, air, fire, and water. Corresponding to these were the bodily "humors," or fluids: blood, phlegm, black bile, and yellow bile. Diseases and disorders were seen as the result of a humoral imbalance. This system—augmented by folk medicine, witchcraft, Christianity, and astrology—had the advantage of completeness. My body and its little world of concerns—toothaches and fevers, lovesickness and moodiness—was part of the wide world and the wider universe. This doesn't mean that the view was that the body was made of the same material as everything else in the universe or that physical forces controlled everything. The ineffable was a genuine and necessary part of reality. Jesus walked on water; miracles happened; the Devil stalked the land. The supernatural—magic—existed within the natural; it was woven into the fabric of the world and the stars, including the sinews of the human body.

At the same time, the system was practical. As a physician in ancient Rome (with a list of clients that ran from Marcus Aurelius to gladiators), Galen himself had favored close observation of the patient—he was the first to recognize the pulse rate as an indicator of health—so that his approach had much to offer it, which explains why it endured for so long. One problem was that the underlying account of the physical world—Aristotle's four elements, which combined in different ways to create all the stuff of reality, from mountains to lily pads to manatees to earwax—did not serve as an especially sturdy foundation. Diagnosis and treatment via the system of humors—a melancholic, or "earthy," illness called for an "airy" compound, and so on—were dodgy if not lethal, as patients well knew and as Molière, for whom the medical profession was a favorite target, suggested with the observation "most men die of their remedies and not of their diseases."

And that was establishment medicine. There were many other options that were considered valid. A sufferer from fever or stomach pain or gout or nosebleed might get, by way of professional service, an astrological reading, an amulet to be tied around the neck with a ribbon, or a squinty examination of his or her urine ("uroscopy" was looked to as a general indicator of health, as when Shakespeare's Falstaff asks a page, "What says the doctor to my water?"). The person administering the attention might be a physician, but astrologers and other sorts of healers were often seen as on a par, and some of the most esteemed medical men, including members of the College of Physicians in London, used astrology as part of their diagnostic tool kit.

Often, the caregiver was a clergyman. In any event, the procedure would have a religious cast. Illness and health were almost universally related to being in or out of God's sight, and the language of healing was shot through with theology. It was commonly held that medicine would work only if a prayer was offered to unlock its powers. Relying on physical remedies alone was often seen as downright ungodly: in England, Puritan minister John Sym advised "caution" that people "*dote not* upon, nor *trust*, or *ascribe too much* to physical means; but that we carefully look and pray to God for a blessing by the warrantable *use* of them." To do otherwise—to rely on a physic or powder alone—would be to put the material above the spiritual. That was why a strictly mechanical approach to medicine was considered dangerously atheistic.

Now, it must be said that millions if not billions of people around the world today subscribe to beliefs similar to those of

Sym: that the physical and the spiritual—pills and prayers, as it were—are both necessary components to health. They visit specialists and get diagnostic screenings, and at the same time they meditate and pray and ask God for a miracle cure. And these people don't exactly inhabit the inner recesses of the rain forest; they live modern lives. They are us. What's more, in the seventeenth century it wasn't only the premodern Aristotelians who held such views; so, for the most part, did the first generation of modern philosopher-scientists who reacted against them. So, too, did Descartes, who seems to have been as devout a Catholic as anyone of his time and whose whole mechanical account of the universe depended on God to hold it in place. The main challenge in following the story of Descartes' bones would seem to be understanding exactly what "modern" is. If it means a hard divide between the material and the spiritual, how do we account for the fact that both people of the seventeenth century who brought the modern sensibility into being and people today have managed to bridge this divide? We associate modern with a nonreligious, nonspiritual, purely rational and scientific outlook. Are we wrong to think that? If so, if it's a false divide, how did it come into being?

A partial answer is that when, in the early seventeenth century, the premodern worldview built around the received wisdom of the Bible and selected ancient writers began to come apart, and as dissatisfaction with it led to a conviction that the mind's latent strength could be brought to bear in radically new ways on the body's weakness, an inevitable result of the new approach was to give greater importance to the physical world and thus, however unintentionally, to devalue theological interpretations. Experimentation was not actually discovered by Francis Bacon in the early seventeenth century, but what Bacon promoted in his *No-vum Organum*, which was published in 1620, was a commitment to reasoning based on observation of the natural world.

The most far-reaching application of this approach came with William Harvey's study of the human heart. Following Galen, the accepted thinking prior to Harvey was that the lungs pumped blood; that there were two kinds of blood, one that was made by the heart and another by the liver; and that both were continually used up by the body. Harvey's dissections and calculations convinced him that the vast quantity of blood that was pumped out of the heart every minute couldn't possibly be consumed by the body. The bold theory he published in 1628—that the blood circulated continually throughout the body, that the heart was the central pump, and that the liver did not make blood—was not instantly adopted by one and all. Harvey anticipated hostility—"I tremble lest I have mankind at large for my enemies. . . . Doctrine once sown strikes deep its root, and respect for antiquity influences all men"—and indeed some medical men ridiculed the idea of grounding scientific work on observation, which, given that the real world was rife with errors and exceptions, seemed patently foolish. Others stayed committed to the notion of two types of blood and thus of the value of bleeding a patient. Phlebotomy (bloodletting) was one of the roots of Galenist healing, and doctors and patients alike clung to it. It was tied both to the theory of humors and to the belief that purging the system was a key to healing, whether it was of the contents of the stomach, the bowels, or a portion of "impure" blood. Observation, however, showed that rather than restoring health, bleeding weakened a patient. To advocates of the new medical philosophy, bloodletting was symbolic of all that was wrong with the old ways—thus the reaction of Descartes, on his deathbed, to the suggestion of it.

Steadily, Harvey's system gained ground in the 1630s; people began to see it as the basis for a whole new approach to medicine, and exploring the recesses of the human body became a fad and an industry and a fascination matching the exploration of the heavens. In Holland, Reinier de Graaf delved into the mystery of birth: he applied his dissecting blade to pregnant rabbits and charted the route the fertilized ovum followed to the uterus. The Dane Nicolaus Steno, working in the hospital of the grand duke of Tuscany, took a step toward demystifying human emotion by laying bare the tear ducts and examining how they functioned. Medical professors created "domestic amphitheaters" in their homes to accommodate the rush of students signing up to observe dissections of human cadavers and vivisections of animals.

In Amsterdam, the physician Nicolaes Tulp gave public anatomy demonstrations, using the corpses of executed criminals. Far from being branded as an atheist, he was immortalized in a painting by Rembrandt in which, using forceps, he pulls aloft a muscle of the left arm of a cadaver. What's more, according to A. C. Masquelet, an orthopedic surgeon who has made a study of the painting, Tulp is holding his own left hand in such a way as to indicate how this particular arm muscle—the flexor digitorum superficialis—controls movement of the hand: the lesson isn't just on the fact of muscles but on the cause-and-effect relationships between parts of the body. The observers in the painting—neatly bearded men with white lace collars—lean in to watch, fascinated by the demonstration. *The Anatomy Lesson of Dr. Nicolaes Tulp* signals one of those telltale shifts in what is deemed socially acceptable—like women wearing pants or the end of segregation in the American South—which to some spell the downfall of civilization while others view the change as an expression of a new era with a new idea of progress. The recesses of the human body, long kept determinedly shrouded in respectful mystery, had become spectacle.

B UT FOR ALL THE interest generated by the great scientific explorers of the late sixteenth and early seventeenth centuries—Galileo, Bacon, Harvey, Kepler, Brahe, and others—their work was fragmented, so that the immediate effect of the endless experimenting, dissecting, peering, and analyzing was more confusion than clarity. Their results didn't fit within the framework of knowledge that had existed for four hundred years. It wasn't possible to use the ancient writers to explain them, and in fact the results threatened to undermine the pillars that had held up the edifice of meaning. It's difficult for us to appreciate what this meant at the time, largely because, as a direct consequence of these men's work, we live in a world with more than one meaning system. Of course, there are fundamentalisms now, too, but even fundamentalists today live with an awareness of relativism. They know there are other systems of belief, even if they are sure those are wrong. In the seventeenth century the challenge to what had been thought an absolute system of values and truths was so sharp and so disorienting that people of all walks of life, from popes to commoners with enough education to read pamphlets decrying the confusion, considered the situation a crisis. And no crisis is deeper than a crisis of belief.

Then, in 1637, a book appeared on the streets of Paris, Rome, Amsterdam, and London. On its title page was an engraved image of a bearded man, dressed in tunic and tights, digging in a garden—the seeker after philosophical truth in the guise of a humble laborer?—above which appeared the full title, written not in Latin but in French so that, its author asserted, it could be read by laypersons (French laypersons, anyway), including, somewhat scandalously, women:

DISCOURS
DE LA METHODE
Pour bien conduire sa raison, & chercher
la verité dans les sciences.

PLUS

LA DIOPTRIQUE.
LES METEORES.

ET

LA GEOMETRIE.
Qui sont des essais de cete METHODE.

Which is to say: *Discourse on the Method of Rightly Conducting the Reason and Seeking Truth in the Sciences. Plus the Dioptric, Meteors, and Geometry, Which Are Essays in This Method.*

The title page also listed the place of publication, the Dutch city of Leiden, and the name of the publisher, Jan Maire, who was at the time unknown but who would become famous because of this one volume. It was printed in an edition of three thousand copies. It would become one of the most influential books of all time.

Conspicuously absent was the name of the writer, who had previously noted that he wished to stay "hidden behind the scene so as to listen to what was said." But the authorship of the *Discourse on the Method,* or *Discourse on Method* as it became known, was identified almost at once.

While he was still at school, Descartes had taken the increasingly apparent faultiness of the foundations for knowledge as a personal crisis. As he writes about it in the *Discourse,* this questioning of values comes off something like the sort of psychological or intellectual crisis that is common in people in their late teens and early twenties: "As soon as I had finished the course of studies which usually admits one to the ranks of the learned . . . I found myself saddled with so many doubts and errors that I seemed to have gained nothing in trying to educate myself unless it was to discover more and more fully how ignorant I was. . . . Nevertheless I had been in one of the most celebrated schools in all of Europe, where I thought there should be wise men if wise men existed anywhere on earth." He cast about for moorings. He wasn't going to be duped by "the promises of an alchemist, the predictions of an astrologer, the impostures of a magician." Regarding the sciences as understood in the Aristotelian system, he judged that "nothing solid could have been built on so insecure a foundation."

Then, like many a college graduate since, he determined to leave books behind and explore the world: "I resolved to seek no other knowledge than that which I might find within myself, or perhaps in the great book of nature." He traveled—for nine years. "I did nothing but roam from one place to another, desirous of being a spectator rather than an actor in the plays exhibited on the theatre of the world." Europe then being caught up in the massive tangle of conflicts known as the Thirty Years War and the Eighty Years War (they ran concurrently), the natural thing for a young man was to learn about the wider world via warfare. He spent time serving in two armies, that of Maurice of Nassau, stadtholder of the Dutch Republic, then that of Maximilian I, elector of Bavaria. He managed to avoid actual combat, however, and instead assisted with solving military engineering problems.

According to Cartesian legend, while he was garrisoned in the Dutch city of Breda he was attracted, one autumn day, by a mathematical puzzle posted on a public notice board. (In an era before newspapers, with their games and diversions, such puzzles were commonly posted in public places.) It was in Dutch, so he asked the young man next to him if he could translate. A fast friendship formed. It happened that Isaac Beeckman had also been distraught over the shaky state of the foundations of the intellectual world in which they had grown up. Both, it seems, had hit on the same new strategy for obtaining genuine knowledge of the natural world: by applying mathematics to physics. The friendship became a kind of competition in which, as Descartes continued to travel with the army, they sent each other problems and investigations, and their correspondence built to a fever of helter-skelter discovery across a dizzying range of topics: music theory, the acceleration of falling bodies, the pressure that fluid exerts on its container, geometry. At the start Beeckman, who was seven years older, was in the role of teacher, but Descartes quickly shot past him, outlining in one letter his discovery of analytic geometry—the use of algebra to analyze geometrical shapes and problems, which in turn would become the basis for calculus—and crowing that once he had worked out the details "there will remain almost nothing else to discover in Geometry. The task is infinite and could not be accomplished by one person. It is as incredible as it is ambitious. But I have seen a certain light in the dark chaos of this science." Modesty was not a condition from which Descartes suffered.

Shortly after, while stationed in Germany, his head teeming with ideas and his whole being straining to comprehend their to-tality, he spent a November night in a "stove"—a tiny room in-tensely heated by a ceramic furnace—and had a series of three momentous, hallucinogenic dreams. On waking, he felt that the dreams constituted a kind of vision: they were the distillation of all the lines of thinking he had been pursuing. The vision was of the natural world as a single system, with mathematics as its key. Pursuing this vision—a new way of seeing the universe and man's relation to it—would be his life's work. Descartes' night of heated dreams has gone down in anecdotal history as one of the fulcrums on which the Western world has turned.

The *Discourse on the Method,* which appeared seventeen years later and summarized the work he had done in that time, was his first published book. To be precise, it was four short books packaged together. The last three were essays devoted to light and optics, geometry and geological and weather phenomena. They included the first or among the first credible accounts of the law of refraction, near- and farsightedness, the nature of wind, cloud formation, and rainbows, as well as the elaboration of analytic geometry.

But it was the introductory essay, the "Discourse on the Method" itself, a mere seventy-eight pages, that gave this small, vain, vindictive, peripatetic, ambitious Frenchman a status among his contemporaries and those who followed unequaled since Aristotle. He was not the greatest mathematician of the seventeenth century (Isaac Newton, a generation later, would surely win that title), or the most influential scientist (here there might be a tie between Newton and Galileo), and one could argue that both Spinoza and Leibniz were more refined philosophers. But Descartes could be considered, as one current philosopher puts it, "the father not just of modern philosophy but, in important respects, of modern culture—of modern Western culture and later, through export of its ideas, of much of modern world culture," and the *Discourse on the Method* is the first reason why. This little essay has been called "the dividing line in the history of thought. Everything that came before it is old; everything that came after it is new."

❦   ❦   ❦

A S FAR BACK AS his school days Descartes had concluded that the place where the traditional approach to knowledge was flawed was at the base: its method of going about the business of understanding. There was no end of brilliance and subtlety in the ancient writers, but if they were starting from a swampy foundation their edifices weren't supportable. Take Aristotle's elemental building blocks of nature: why earth, air, water, and fire? What is the rationale for supposing that what is simply most evident to the senses is necessarily the base of reality? Or consider Thomas Aquinas, the finest of the Scholastic thinkers, who devoted his razor intellect to such things as an elaborate "proof" of the existence of angels, which included an analysis of their numbers, varieties, substance, intelligence, and origin and resolved such questions as whether, in moving from one place to another, an angel passes through intermediate space. How could one of the greatest minds in history get itself into such convoluted alleys of reasoning? Or Plato, with his theory of forms, according to which the tree out the window is not itself real but merely a reflection of the eternal form "tree" and the keyboard on which I type is actually an imperfect approximation of a perfect nonmaterial form—call it "keyboardness"—that was created by God and exists in eternity.

Layers of tradition had built up around such categories for understanding reality. Centuries of robed scholars and scribes had bent in tallow-tapered light over parchment sheets and leather-bound manuscripts, mouthing words, quill-scratching, rubricating, memorizing, parsing and analyzing and adding levels to the hoary infrastructure that had these categories as elements and that was applied as an increasingly unwieldy tool to explain natural phenomena, human behavior, history, the universe. But on what ground did they stand, these classifications? How could one trust them? How do we know they aren't nonsense? Or, if they were true, couldn't we expect that great things would have arisen from knowledge built upon such bases? As Descartes put it, devastatingly, "The best way of proving the falsity of Aristotle's principles is to point out that they have not enabled any progress to be made in all the many centuries in which they have been followed."

What kind of method, then, would yield progress? Descartes was clear as to his ultimate aim. Unlike philosophers of later eras, who would devote themselves to questions of the order of "Why is there something rather than nothing?" he was full-blooded in his inquiry: he was after the kind of philosophy that would take the world by the throat, that would make people "the lords and masters of nature."

At first glance, his way of proceeding doesn't seem to make sense. For a universal measure one might reasonably look outward, like a navigator with a sextant: to the stars, to the distant horizon. Instead, his break from tradition is signaled, first, stylistically: the *Discourse on the Method* is written in the first person. A byproduct is that one of the world's great works of philosophy is also one of the most readable. And it serves as an appropriate launching point for a new era in which the focus is on the individual. The *Discourse* begins not with mathematical formulas or scientific propositions, not with the lining up of outside authorities, but with a living human being—Descartes himself—sitting alone, thinking. There is atmosphere in the text, snugness: you can almost hear the crackling of the fire in the background. The realm we're in is familiar: it's that of the novel, the narrative, the play, and the film. It's human and, yes, modern.

All of these modern art forms involve, in addition to a personal focus, a central crisis on which the story turns, and so does this first work of modern philosophy. The crisis is a loss of meaning, and the quest is a search for truth, for something to believe in. Descartes' strategy was to assume that Aristotle's entire approach to nature, to reality, is wrong and then to assume the same for Aquinas, Plato, Duns Scotus, William of Occam, and all the revered writers. He ceremonially placed the Bible—from Adam and Eve to the Hebrew prophets to the resurrection of Jesus Christ—in the same dustbin. He continued slashing every such thought and idea until he came to a proposition that was impossible to deny. It was both a philosophical and a psychological undertaking, and to it he appended a "don't try this at home" caution: "The single design to strip one's self of all past beliefs is one that ought not to be taken by every one."

Maybe grand abstract writings needed to be dismissed in this way, but what about the things that are right in front of me? What about, as Descartes put it in *Meditations on the First Philosophy*, the simple fact "that I am in this place, seated by the fire, clothed in a winter dressing-gown, that I hold in my hands this piece of paper"? Even these things fell by the wayside. The senses can't be trusted either. The senses deceive. I might be dreaming, or drugged, or deceived by a malicious deity. If we are being serious about this project, then sights and smells and tastes, no matter how self-evident, must also be doubted. Strictly speaking, I can't even be sure of the reality of my own body.

Which leaves what? At the end of this remorseless reduction there is only one thing that remains, one proposition that can't be denied, one sound, as it were, in the universe, like the lonely ticking of a clock. It is the sound of the thinker's own thoughts. For can I doubt that thoughts are occurring right now, including this one? No: it's not logically possible. So, humble though it is, we can call this a ground: bedrock.

In this way, Descartes became one of those rare figures in history who have given the world a sentence that is a touchstone. "I am the way, the truth, and the life" was such an utterance, standing on one side of Descartes and his era. On the other side we have "E=mc^2." As philosophers since have pointed out, "I think, therefore I am," or "Je pense, donc je suis," or "Cogito, ergo sum," does not fully encompass what Descartes intended. Once the acid of his methodological doubt had eaten its way through everything else, what he was left with was not, technically, even an "I" but merely the realization that there was thinking going on. More correct than "I think, therefore I am" would be "Thinking is taking place, therefore there must be that which thinks." But that hardly has the snap to make it a slogan fit for generations of T-shirts and cartoon panels.

The irony is that in shifting the focus onto the individual human mind, which everyone agrees can be a pretty flimsy and wayward organ, Descartes had arrived at the closest thing to a certain basis for knowledge. If my own thoughts are the only indubitable ground I can stand on, apparently they aren't so flimsy after all, at least not all the time. As an early follower of Descartes put it, "Doubt is the beginning of an undoubtable philosophy." Therefore the mind and its "good sense"—that is to say, human reason—are the only basis for judging whether a thing is true. With the "cogito," as philosophers abbreviate it, and with the theory of knowledge that arises from it, which Descartes outlined in the *Discourse on the Method* and later works, human reason supplanted received wisdom. Once Descartes had established the base, he and others could rebuild the edifice of knowledge. But it would be different from what it had been. Everything would be different.

T HE MASSIVE GOTHIC CHURCH tower that dominates the skyline of Utrecht stands oddly alone in the middle of a central square of that Dutch city today, separated from its cathedral by a wide swath of paving stones. The explanation for this anomaly is that the nave that connected the two structures collapsed during a violent storm in 1674 and was never rebuilt. With this exception, the exquisitely preserved old city center, with its sunken canals, twisty streets, and gabled brick façades, is not so different from the way it would have appeared in the year 1638, when a bluff forty-year-old medical doctor named Hendrik de Roy—who was known by the Latinized form of his name, Regius—swept determinedly into the building connected to the cathedral, the newly christened Utrecht University, to take up his duties as a professor.

Shortly after, Regius penned a letter to René Descartes, whom he had never met but who was then living in the village of Santpoort, forty miles away (for most of his career Descartes lived in the Dutch provinces, to which he was attracted by the comparatively greater atmosphere of intellectual freedom). Regius wanted to thank Descartes, for, he said, he owed his appointment to the newly created chair in medicine to the Frenchman. Regius had read the *Discourse on the Method* when it came out the year before, including the accompanying essays on optics and meteorology, and the book had changed his world. He had previously been teaching private lessons in physics; after reading the *Discourse* he revamped his whole approach, and now his lectures were packed with students intrigued by this new way of understanding the body and, for that matter, the universe. The regents of the university took note; Regius believed that the popularity of the courses had led to his promotion.

Regius asked Descartes to accept him as his "disciple." Descartes was delighted: he was highly susceptible to flattery, and besides that this was the type of response he needed if his work was to have an impact. Although initial sales of the *Discourse* were modest (sounding a note in harmony with authors of all eras, Descartes whined that the print run wasn't selling out, so he doubted there would be reprints), the book was being read, and he was in the process of moving from obscurity to intellectual celebrity. His name was being uttered in universities, churches, taverns, and cafés. The chattering classes of seventeenth-century Europe were writing to one another and to him using shameless superlatives, referring to him as "this great man," "the Archimedes of our time," "greatest of all philosophers," and "mighty Atlas, who supports the heavenly firmament, not with raised shoulders, but by the firm reasoning of [his] magnificent mind." The interest was in, as Regius put it, "this excellent method" that Descartes had discovered "for conducting reason in the search for all sorts of truths."

Descartes was by now formulating his goal in life, which was nothing less than to supplant Aristotle as the basis for education. It was an almost laughably grandiose ambition, far greater than that of Galileo or Leonardo or even Aristotle himself, who after all did not think he was laying the intellectual framework for centuries of human history. Not to put too fine a point on it, Descartes wanted to reorient the way every human being thought.

And he knew you didn't do that by writing a book or two and giving some lectures. If his philosophy was to be adopted he had to build a network: to win over influential professors, church officials, university overseers, and government leaders. This process would start with Regius.

Regius was keen to promote Cartesianism; with Descartes' blessing, he gave a series of formal disputations on the subject at the university. Since he was a professor of medicine, the theme was "the science of health"—a title that in itself denoted the clinical approach he would present. He seems to have built up the drama, and the audience, as he went: each talk was progressively bolder. Starting from the radical new ground of individual human consciousness, which is able to manipulate the tool of reason, he laid out the Cartesian universe in all its machinelike regularity: from the Copernican view of the heavens, in which the earth is just one of many bodies revolving around just one of many suns, to the organs of the body, which depend not on a divine agency but on the proper physical conditions for their functioning. He delved into details: the action of the pulse, respiration, and even "de excrementis." He largely ignored theology, openly ridiculed the Aristotelian categories, and presented Harvey's theory of the circulation of the blood—which Descartes had taken up and made a central part of his own physiology—as the key to the body's functioning. (While Descartes followed Harvey's overall theory of circulation, he disagreed with Harvey's thesis that the heart pumped blood. Descartes believed—wrongly, of course— that the heart acted like a furnace that heated the blood, causing it to circulate.)

But while some intellectuals across Europe were poring over the *Discourse* and crediting it with creating a new framework for knowledge, the response to Cartesianism at Utrecht—arguably the first public reaction to the idea that we call modern—was different. At the end of Regius's last talk, the packed room erupted in catcalls and chaos. The dignitaries stormed out. The mayhem was unprecedented, and a full-scale crisis was under way.

For all his ambition, Descartes didn't like direct confrontation, and he had kept away from Utrecht during this time. Regius, however, with his boxer's face and querulous disposition, was all for mixing it up with the Aristotelians, which further inflamed the crisis. When a physician named Jacques Primerose challenged him on the theory of the circulation of the blood, Regius fired off an essay that he called "A Sponge to Wash Away the Filth of the Remarks Published by Dr. Primerose."

The real enemy, however, was Gysbert Voetius, who in addition to being a theologian and an Aristotelian was also the rector of the university. If Regius had recognized at once the sweeping promise of Descartes' philosophy, Voetius—a small, intense, ferret-faced man—immediately saw the danger in it. Voetius mounted an attack on Regius, Descartes, and the "new philosophy," charging that Copernican astronomy, which Descartes had taken as part of his system, was an affront to "sacred physics" that was creeping insidiously into the minds of the European intelligentsia and had to be eradicated. More to the point, Descartes had promised that his philosophy, grounded in his method, would bring sweeping new insights into nature—but according to Voetius this level of knowledge was not of this world but spiritual: it was of the order of the "Kingdom of Heaven" of which Jesus taught. "There is so much we do not know!" Voetius declaimed. Given that fact, the way to truth was not via the destructive doubting of all that had been carefully built up over the centuries but through cultivating and reverencing a "learned ignorance."

Voetius used his influence to wage his campaign against Cartesianism on several fronts. He warned that it was a road leading to the most dangerous of places the human mind could go: atheism. He and his supporters accused Descartes of being something akin to a cult leader, who kept his followers in thrall by means of personal magnetism. That was the real purpose of Cartesian doubt: by encouraging his followers to forget what they had learned from the ancient masters, Descartes was emptying their minds, so that he could fill them with his own teachings. Then, for good measure, and in the time-honored fashion of smear campaigns everywhere, Voetius and his allies accused Descartes of sexual deviance.

Descartes was stunned by this reaction to the introduction of his philosophy. He may have shied away from face-to-face confrontation, but his arrogance was rather spectacular, and when crossed he had a deep malicious streak. As one opponent wrote, "It may be true that Descartes tries to free himself of all prejudices, yet there is one to which he remains singularly attached— the conviction that he is absolutely right on everything." When it came to dealing with criticism, Descartes could be thin-skinned and downright scatological. The mathematics of Fermat he once characterized as "shit"; another prominent writer's work he described as toilet paper. In an argument with Pascal over the existence of a vacuum in nature (Descartes insisted there was no such thing, Pascal said there was), he tossed off the witticism that the only vacuum was in Pascal's skull. When the second edition of his *Meditations* came out, he took the opportunity to add in an appendix his version of the events in Utrecht. Not content to leave it at that, he then wrote a two-hundred-page response, which he published under the title *Letter to Voetius,* defending his philosophy and accusing the theologian of being a demagogue, a slanderer, and, turning the tables, an atheist. On top of which he achieved a rhetorical pirouette in making a case that the real reason for the vitriol thrown at him was that Voetius and his cohorts knew that Descartes was onto the truth and they couldn't face the reality of their worldview's collapsing. The effect of all of this was to bring the whole affair to the wider European audience.

For his pains, Descartes found himself charged with libel and facing the possibility of a criminal sentence. He called on the French ambassador for help, who in turn went to the stadtholder, the chief magistrate of the province. As the affair wound through the Dutch system, it became something of a public event: town officials, theologians, regents, clergymen, magistrates, professors, and students pondered the issues involved, weighing whether the new philosophy, now fully tricked out with its Cartesian underpinnings, was a genuine path to knowledge or an assault on Christianity or—perish the thought—both. There was a palpable feeling that the foundations of society were under attack. The lawsuit reached one climax in 1642, when the city of Utrecht formally banned Descartes' philosophy.

The controversy didn't stop there: the panic and confusion spread. In 1647, the debate reached Leiden University, the most prestigious institution of learning in the country and the very city where the *Discourse on the Method* had been published. The university, like Utrecht before it, tried to settle the matter by banning Descartes' philosophy, decreeing that "neither in public nor in private lessons must philosophers depart from Aristotle's philosophy." As in Utrecht, Cartesianism was debated in a public disputation that became violent. People stood on tables and benches, packing the auditorium so that "heads reached to the vault." A philosophy professor named Adam Stuart presented the case against Descartes: "There are certain newfangled philosophers who deny that the senses can in any way be trusted and claim that philosophers can deny that there is a God or that one can doubt His existence." By now there were quite a few disciples of Descartes' among the faculty and students, and some expressed outrage that Stuart was misrepresenting him. Tempers rose; students stamped their feet and hooted. Stuart came completely unglued and took to shrieking at the principal defender of Descartes, "In virtue of my public authority, I order you to be silent! Shut up, I do not want to hear you!" As in Utrecht, the event ended in chaos.

What was at issue in these confrontations—which would be repeated over and over across Europe in the ensuing decades—was the nature of the relationship between faith and reason and also the relationship between the spiritual and physical worlds. In each, according to tradition and law, the former had precedence over the latter. Then again, the battle lines between faith and reason have never been clear-cut. Descartes himself was not the cool rationalist that history has portrayed him as. He held sincerely to his faith, and while he was undeniably a modern philosopher, he also had one foot in the Middle Ages. In the manner of medieval philosophers, he incorporated "proofs" of the existence of God into his philosophy. It was necessary for him to prove both the existence and the innate goodness of God, for, given the corrosiveness of Cartesian doubt, these were the only assurances we have that the material world really exists. So his work has a theological grounding: not only do the world and science depend on God, but so does Cartesian philosophy.

The Aristotelians weren't buying it. Descartes' method used reason as its tool, so its approach to the physical world downgraded theological readings. To pursue it as an actual basis of thought, they argued, would lead to atheism, to a breakdown of authority, to a world riven by doubts and confusions, without any arbiter, any rules.

What is most remarkable in all of this is how familiar it sounds to twenty-first-century ears. There are a lot of people today who would say that this is precisely where we are now: rudderless, suffering the consequences of relativism and doubt. As Joseph Cardinal Ratzinger put it just before he became Pope Benedict XVI, "We are building a dictatorship of relativism that does not recognize anything as definitive and whose ultimate goal consists solely of one's own ego and desires." Much of what he and other current religious conservatives say sounds identical to the fears of a contemporary critic of Descartes who warned that Cartesian "doubts might be imported from philosophy to theology. Students would go on to doubt everything: themselves, God, et cetera." On the other side are those—secularists or religious moderates—who insist that blind faith is the problem that humanity has to overcome. The striking thing is that within a matter of months of the publication of the *Discourse on the Method*, the first work of modern philosophy, the crisis that might be said to define the modern era—the deep and complex confusion over the realms of faith and reason, which would appear again and again in different guises in the coming centuries, from the French Revolution to Darwin's theory of evolution to present-day debates on how to deal with militant theocracies and religion-fueled terrorism—had already appeared.

Occasionally, it's not only a philosophical inheritance but a concrete event from the dawn of modernity that sprouts into the present, like a seed from an heirloom plant in a field of bioengineered crops. In June of 2006, I was invited to lunch at the faculty club of Utrecht University by Theo Verbeek, one of the most distinguished Descartes scholars in the world, and Erik-Jan Bos, a philosopher at the university. Together they are editing a critical edition of the correspondence of Descartes (which runs to five

volumes—in his zeal to promote his philosophy Descartes was a
formidable communicator). Verbeek told me of a remarkable event
he had participated in the year before. In a formal ceremony (with
a Latin text, no less), officials of the city and the university issued
a public apology to Descartes for the treatment he had received,
lifted the ban of 1642, and, in the words of *Le Monde,* which
covered the event, "solemnly rehabilitated" the philosopher. The
event made news as far away as Japan. In 2000, Pope John Paul II
had apologized for the Catholic Church's having put Galileo on
trial over his Copernican views. "We modeled it on the Vatican's
ceremony," Verbeek said to me. "Utrecht was the first place in
the world to recognize Cartesianism and the first place to ban it.
We've finally corrected that."

During our lunch we also talked about the house Descartes
had lived in briefly in Utrecht, which has long since vanished. As
we left the dining room, I asked Verbeek whether there were any
landmarks particularly associated with Descartes' time in the city
that were still standing. He gave me a slightly incredulous look—
as if to say, "You mean you don't know?"—then led me around a
corner and opened a large wooden door. The room it gave onto
was a vast, sweeping, churchlike space with high vaulted ceilings
and ornate wood carving. A doctoral dissertation was in progress,
which in the Dutch system is a formal affair, requiring scholars
to wear academic robes and mortar boards—giving an extra turn
of historical verisimilitude to the scene. For me, it was an electric
moment, all the more so because I hadn't been expecting it. This
was the aula, or auditorium, of the university, unchanged in more
than three centuries. It was where, in 1641, Regius had given his
disputations in support of Descartes' ideas—the place, you might
say, where modernity first went public.

❖   ❖   ❖

I N THE YEARS AFTER he hit on his "method for rightly conducting the reason," Descartes applied it obsessively, voraciously, to everything he could think of. Ironically for a historical figure who is remembered as the ultimate conceptual thinker, whose very name has become an adjective paired with abstractions (Cartesian coordinates, Cartesian dualism), he was tremendously fixated on the mundane. He studied snow, rocks, grains of salt. He was fascinated by the idea of bringing his method to bear in law and once took on the case of a peasant who had been accused of murder. He investigated the details and appealed to the authorities on the man's behalf: a use of reason that presaged Sherlock Holmes and forensic science.

These pursuits reflected his main motivation, which he shared with contemporaries like Bacon and Harvey and which we have inherited from them: the idea of human progress. "Mastery of nature," Descartes believed, would lead to "freedom," and by this he meant freedom from drudgery, freedom from prejudices and errors of thought, and of course freedom from pain and disease. For the sickly child shivering in fear of death never went away: Descartes kept human health as a chief focus. When he lived in Amsterdam in 1635, his address on the Kalverstraat was a convenient one: the street was named for the calves and oxen that were slaughtered by butchers there. He had to walk only a few steps— along a row of two- and three-story gabled houses, in the shadow of the baroque clock tower called the Munt—to find the freshly killed specimens that he would cart back to his home, where he would slice them open to search the mystery of the eyeball, the dark tangle of the intestines, the chambers of the heart. In his years of medical research he dissected rabbits, dogs, eels, cows. As in other areas, his self-confidence, not to say arrogance, was vast. "I doubt whether there is any doctor who has made such detailed observations as I," he wrote to a friend. "But I have found nothing whose formation seems inexplicable by natural causes. I can explain it all in detail."

Elsewhere he noted blithely, "Now I am dissecting the heads of different animals in order to explain what imagination, memory etc., consist of." If the arrogance was uniquely his, this sort of naïveté with regard to the imminent payoff of science was something he shared with his contemporaries. There was a tendency, in that idealistic moment at the birth of the modern, to believe that solutions to some of the most complex of human problems were just around the corner, and Descartes was convinced that his method was the key. He insisted, for one example, that to achieve mental health one need only turn the method of reason on oneself. Once people did that, perversions and manias would fall away, as would ordinary feelings of jealousy, fear, greed, and anger.

The optimism went further still. The body was a machine; therefore it simply needed to be understood in all its parts in order for it to work properly. In this regard, death was tantamount to a malfunction; locate and correct the errors and you solve the problem of death. Descartes became convinced that he would crack the body's code and extend the human life span to as much as a thousand years. At one point in his career he was certain enough of his progress that he felt he would do it soon, provided, he wrote—and he seems to have missed the joke—that he were not "prevented by the brevity of life."

As absurd as Descartes' hopeful ideas about medicine seem today, they were believed by others. Europe's greatest intellectuals were as aware of his medical focus as they were of his philosophi-

cal and mathematical discoveries. He gave Blaise Pascal medical advice. The Dutch statesman and poet Constantijn Huygens
wrote to ask if Descartes could pause in his investigations long
enough to give a few hints on "the means to live longer than we
now do." Royals wanted to be updated on his progress, to know if
he had yet found a way to increase the life span "to equal that of
the Patriarchs." He seemed to relish the role of medical adviser,
but in the absence of some basic medical discoveries—germ theory, blood types, anesthesia, microbes, and bacteria were centuries in the future—much of his advice was sound if unexciting: he
tended to recommend rest, soup, and keeping a positive outlook.
Meanwhile, proving that he was truly modern, as he saw gray
hairs sprouting on his head after he turned forty, he got to work
on a process he believed would retard graying.

Of course, Descartes did not go down in history as a medical researcher. His focus broadened eventually beyond the human
body, and one reason for that broadening had to do with the body
itself and its frailty. Descartes never married and had almost no
known intimate relations, but he was, after all, human. We happen to know that on October 15, 1634, while he was lodging at
the home of his associate Thomas Sergeant in Amsterdam, Descartes had sex with a woman named Helena Jans, who was his
host's housekeeper. We can be precise about this because nine
months later a child was born, and afterward Descartes gossiped
to a friend and told him the date of conception, of which he had
obviously made a note. Helena Jans was in the employ of Sergeant, an Englishman who ran a bookshop and who lived in a
house that still stands in the center of the city (it is just around the
corner from the Anne Frank House), but when several months
later Descartes left Amsterdam for the Dutch town of Deventer,

where he had previously been living, Helena Jans went with him. Her daughter was born there; they named her Francine.

Descartes was a self-centered, vainglorious, vindictive man who stayed remote from his family and had few close personal friends, but with the arrival of the little girl he seems to have changed in some basic way. The child was, according to his early biographer Adrien Baillet, the love of his life. Fathering a child out of wedlock was a serious offense, and Descartes took pains to keep the fact hidden, but at the same time he acknowledged the baby as his own in the baptism registry (though he used only his first name).

They moved around over the next few years, this curious little family, Helena officially as his servant, Francine as his "niece." In 1640, he wrote to a female relative in France, saying that he was preparing to bring his daughter there to learn the language and be educated. In early September, before the move to France, he took a short trip to Leiden. Then came news: Francine had been stricken with scarlet fever, "her body all covered in purple." According to one account, he returned in time to hold her in his arms as she died. She was five years old.

The five years of his daughter's life, followed by her death, formed a kind of pivot around which Descartes' work turned. Before, a main focus was on medicine and curing disease, which came in part as a result of his own childhood illness. He dissected animals, as if expecting to find, somewhere within the recesses of the body, an actual key. The experience of fatherhood, and then of losing a child, coincided with a broadening of his focus. It was as though staring into what is surely the blackest of all holes—the grave of one's child—pulled him out of the body, led him to conclude that it would not give up its secrets so easily, and compelled him to look to the universe for answers.

❧   ❧   ❧

A T THE SAME TIME, however, Descartes maintained confidence in science and its power to aid and extend human
life. And that, surely, was one component of the anger that wracked
him as he lay dying on an icy February night in Stockholm. And
with the anger came bitterness, for he hadn't really wanted to
come here to begin with. Stockholm was far from the European
centers of power. Besides that, he detested the cold. He had been
born and raised in the Loire Valley, the sunny garden of west-
central France, and he cared about personal comfort. Sweden he
dismissed as a remote "land of bears between rocks and ice."

But the stream of letters from his friend Chanut inviting and
urging him to Stockholm had come as he was at a low ebb. He had
been living in the Dutch provinces for more than twenty years,
working to get his philosophy accepted in what were supposedly
Europe's most freethinking universities, but the battles in Utrecht
and Leiden had worn him down. He had begun to realize the difficulty of the task. He felt old and tired; a change was in order.

Then, too, he had another reason for going north: the queen
of Sweden. It was probably partly because of her that he accepted
Chanut's invitation. It was no doubt due to her that he, normally
a sober dresser, did himself up in peacock fashion—long, pointy
shoes, gloves of white fur, and his hair specially curled—as he
boarded the ship that came expressly to pick him up on the Dutch
coast.

He met her the day after his arrival in Stockholm. He may
have been surprised at first. Queen Christina did not, even according to her admirers, cut a dazzling figure. She was plain, with
a prominent nose and mournful eyes. She didn't care much for

dressing up if she didn't have to (she allowed all of fifteen minutes to ready herself in the morning). She was short and a bit stocky, with tiny feet. She liked riding, hunting, and shooting, and when she was on horseback she wore a man's collar, so that Chanut, in describing her, once said that anyone who didn't know the rider was the queen of Sweden would think her a man.

This was one of many hints that would lead later writers to claim that the doctor who had delivered her had spotted a tiny penis; there were rumors of gender bending, hermaphroditism. Probably they were the sort of sexual slander spread by men who didn't know what to make of a woman who wielded power, but during her prime she managed to behave in such a way as to ignite gossip. She never married, for one thing, and once remarked with loathing that being a wife would be like being worked over "as a peasant does with his field." She was said to treat certain young ladies with unusual familiarity and, with one in particular, to "perform immoral acts." Meanwhile, she allowed a French physician at her court free rein over her bedroom—he wrote lasciviously to a confidante that she had "begun to taste," whatever that means.

But despite appearances, all agreed that when she opened her mouth to speak everything changed. Christina's personality—crafted by the circumstances of her life as well as by her intellect and learning—was unstoppable. She was brilliant, intellectually voracious, and domineering. She was only twenty-two when Descartes arrived at her court, but people had been talking about her for many years—since her father had died in battle in 1632, making her queen at age six. She grew into a psychologically complex monarch, maybe too much so for a stolid country to know what to do with. Looking back on her early childhood must have been like mulling a hazy dream. Her father, King Gustav Adolf, had held sway over a land that was still medieval, a nation made up of scattered small farming settlements dotting an endless expanse of meadow and pine and birch forests. He was a mythic figure, blond and Nordic, a dreaded warrior whom enemies called "the lion of the north" and who held court in the open air; his peasants paid taxes to him in kind—cattle, barley, oats, hides. Currency existed in his time, but the daler, because it was of copper, the country's chief metal, tended to be unwieldy: a coin was typically the size of a dinner plate.

Gustav Adolf wanted Sweden to become a power to rival the nations to the south, and he largely succeeded, thanks to his skill in maneuvering Swedish warriors through the thicket of the Thirty Years War. By the time he had fallen in the mud of lower Saxony, with bullets in his back, arm, and skull, Sweden commanded respect from its neighbors and was moving toward a sophisticated society, economy, and governmental bureaucracy. His chancellor, Axel Oxenstierna, held the reins of power until the young Christina reached maturity. Like Alexander the Great, she was groomed from early life in books and the art of war. One of the things that had intrigued Descartes was the reputation she had as a ferocious intellect, someone who could keep up with even a great philosopher. Building on her father's achievements, she now wanted to turn her court into one that would rival that of France. She wanted artists, poets, philosophers. She wanted an academy of science. Thus the invitation to Descartes.

But, too, on her side there was something deeper. If the force called modern asserted itself most clearly in science and philosophy, there was also a political dimension. The year before Descartes arrived in Sweden—1648—represents one of history's political watersheds. The Thirty Years War and the Eighty Years War—which together had engulfed most of Europe—came to an end simultaneously in that year. The treaties that ended the two wars were seen by those involved as a dividing line between past and future. If you were alive then, you would have taken part not in a single celebration but in an ongoing party that spread across the months and into the next couple of years. (Not long after Descartes arrived in Stockholm, Christina sponsored a ballet—*The Birth of Peace*—devoted to the topic. Its libretto was long believed to have been written by Descartes, on the queen's orders, but recently the philosopher Richard Watson has argued convincingly that Descartes was not the writer.) Those protracted conflicts, largely religious in nature, involved a level of slaughter that had never been known before. The negotiators charged to end them crafted, in the process, a new secular sense of the relations between nations and, with it, new thinking about the meaning of peace and how societies might relate to one another. Rather than being tied to Rome, or to one of the Protestant confessional blocs, nations saw themselves as independent actors that could control their own fates. It was, in a word, the advent of secular politics.

In this new age, the leaders of the nation-states of western Europe were hunting for innovative tools and tactics that they could leverage to their advantage. Science—or Cartesianism, or new philosophy—was coming to the fore at the same time this political thinking was evolving, and political and military leaders looked to it as a potential source of power. In a sense it was a trope that had played out endlessly before and would continue to do so in the future. The duke of Milan had hired Leonardo da Vinci to create military hardware; the United States would secretly smuggle Wernher von Braun and other German rocket scientists out of Germany during and after World War II, scrub them of their Nazi associations, and put them to work founding the American space program. Christina got reports on the frenzy of scientific exploration going on across the continent—people doing unheard-of things with cadavers and flower bulbs, injecting quivering animals, gazing at the heavens, predicting imminent findings that would shake society to its foundations—and she wanted to be a part of it.

After his first, hopeful meeting with the queen, Descartes took a floor in Chanut's house and tried to settle into life at court. He quickly found, however, that the other intellectuals she had assembled resented him. He also discovered that, where earlier Christina had been keen on his philosophy—she had read his latest book and had written to him herself, through Chanut, posing questions on the nature of love and on how the modern notion of an infinite universe could be squared with Christian belief—she now seemed to have moved on to other things.

In particular, she had fallen heavily for Greek esoteric knowledge, a semimystical inquiry into nature that relied on ancient writers and that for a brief time rivaled the mechanistic new philosophy as a potential replacement for Aristotelianism. When Descartes discovered that the queen was giving much of her attention to the study of ancient Greek, he reacted as if it were an illness she had caught, saying to a friend, "Perhaps this will pass." He, after all, wanted to sweep away the old learning in favor of science and experimentation and considered such study a colossal waste of time. He soon came to see Christina as a dabbler and a dilettante. And she seems to have been disappointed in him, too: he appeared more doddering curmudgeon than fiery revolutionary. His philosophy didn't seem transferable into political power or, for that matter, personal growth.

The mutual disillusionment played out over the course of the winter, and in circumstances unfavorable to Descartes. He liked to work at night and sleep in late; she always woke at four in the

morning, and she decreed that he would give her philosophy lessons beginning at five o'clock. In the black predawn he lumbered
by coach from Chanut's house over the little hump that was the
center of the island that formed Stockholm's core and trudged
up to the castle perched grandly above the harbor. It was cold,
the coldest winter in living memory, his lifelong fear of catching colds and fevers reasserted itself, and he became dark. "Here
men's thoughts freeze like water," he wrote in the last letter of his
life. And added frankly, "I am out of my element."

And so came illness, and its worsening, and then the realization—after calling for remedies of his own concoction (for example, wine infused with tobacco to induce vomiting)—that he
would not recover. Outside, far away, forces that he had helped
set in motion were continuing without him. Letters came from
Paris, London, Amsterdam. Pascal wanted updates on an experiment on barometric pressure for which Descartes had been giving him temperature readings. Who knew if a month or a year
hence wouldn't bring a discovery related to the regeneration of
human tissue or proof that celestial bodies were governed by the
same forces as those on earth, which would firmly establish his
mechanical notion of the universe? Walls were collapsing, scales
were falling from people's eyes. And here he lay, in this remote,
cold, stony world, a veritable tomb.

Finally Descartes agreed to let the physician Wullens see him
again. But he remained cantankerous, so that Wullens was baleful
in his Latin follow-up report, calling his patient *homo obstinatus*
and complaining afterward that Descartes had told him "that if
he had to die, he would die with more contentment if he did not
have to see me."

Then came the final indignity, which Descartes not only relented to but, in ultimate capitulation, asked for out of the ex-

tremity of his desperation: to be bled. Three times his arm was opened; the blood that pulsed out, Chanut's secretary noted, was "oily." Rather than improvement, Wullens reported, there came "the death-rattle, black sputum, uncertain breathing, wandering eyes." When death arrived, it was seemingly with spite on its breath.

N OW THIS WAS AWKWARD. Chanut and Christina had lured the great man to them, they had taken him under their protection, and then they had, well, killed him. Christina could let his death wash off her royal personage, but Chanut, as both friend and diplomat, felt the full brunt of guilt. Much as he would have liked to avoid them, there were responsibilities to perform: he had to break the news. His letters fanned out across Europe, beginning the day after the death. To the comte de Loménie, the former French secretary of state, he lamented: "We are afflicted in this house by the death of M. Descartes, . . . a rare man of the century." Perhaps to deflect responsibility, he went on to explain that the philosopher had been in Sweden because "the Queen of Sweden had desired to see him with a passion." To Princess Elizabeth of Bohemia, who had become close friends with Descartes, Chanut bowed low in anguish: "I . . . say to you, Madame, with an incredible pain, that we have lost Monsieur Descartes." While to Claude Saumaise, a French linguist whom Christina had invited to her court, he wrote, "Mr. Descartes, who gave us the method and the design, will not have the pleasure to see the beginning of it"—meaning, presumably, that Descartes would not live to see the full flowering of science. And here Chanut tried to assuage his guilt a bit by noting that Descartes had died after an illness "in which he did not want to avail himself of the assistance of the Doctors."

The news of the death spread and, oddly enough, caused some bewilderment. The idea that Descartes would end disease and dramatically lengthen life had become so widely held in certain circles that some intellectuals refused to believe he could be dead. "Impossible," wrote the French abbé Claude Picot, who said he had been convinced that Descartes "would have lived five hundred years, after having found the art of living several centuries." How could the chief investigator of longevity die so young? There had to be a sinister explanation, something that, as Picot said, "deranged his machine." A rumor sprang up, which circulated for decades, that he had been poisoned.

Meanwhile, there was the matter of the body. Christina announced that she wanted to bury the great philosopher in Stockholm. If in life he hadn't been the ornament to her court that she had wanted, perhaps in death he could add some luster. Chanut's position would logically have impelled him to insist that the body be returned to France, but then again that would draw another, greater round of attention to the awkward fact that the man had died here, under his watch. He acquiesced.

But where to bury him? There was no question in Christina's mind. He would be given a full ceremony and laid to rest in the Riddarholm Church, the ancient resting place of Swedish kings, whose number included her father. Chanut was appreciative of the high honor offered to a countryman, but he definitely did not want this and set himself obsequiously to convincing the queen to rethink her decision. He sent his secretary, an especially pious man named Belin, to the castle to explain that his reasons for desiring other arrangements had to do with religion. Descartes was a Catholic; France was a Catholic nation, which wouldn't ap-

preciate one of its native sons being buried in a Lutheran setting. Plus, there had already been whispers at Christina's court
that Descartes and Chanut together were trying to convert the
queen from Swedish Lutheranism. Perhaps Her Majesty could
appreciate, Belin offered, that burial in the state church might
be . . . undiplomatic?

Christina relented, and if Chanut's deep objective was to keep
the whole thing as quiet as possible, he achieved it. The place,
time, and circumstances of burial might have served for a plague
victim. At four o'clock on a winter morning, barely twenty-four
hours after the death, a small procession traveled a mile north of
the center of Stockholm, wagon wheels creaking in the frozen
ruts, and turned into a lonely little cemetery whose charges were
mostly orphans. Apparently Chanut had made some inquiries and
concluded that because children who had not attained "the age of
reason" were not considered to be outside the graces of the Catholic Church, such burial ground, if not exactly sanctified, could
not be said to be unholy. Theologically, it would do. Better still,
it was remote.

Four men—one of them Chanut's seventeen-year-old son—
carried the coffin to the waiting grave. A small group of people
gathered around it in the frozen darkness, their faces lit by flick-
ering torches. Beneath the icy swirl of the indifferent heavens, the
sole priest invoked the name of God. Dirt skittered on the coffin
lid. Then everyone went home.

Banquet of Bones

"DEATH," THE PHILOSOPHER LUDWIG WITTGEN-stein once wrote, "is not an event in life." He meant, maybe (for it's hard to be sure—Wittgenstein was rather titanically cryptic), that being dead is not something we actually experience and that since we aren't conscious of a nonliving state it is literally meaningless, so instead of spending our lives worrying about the future we should look at each instant as an eternity. We should live in the moment.

Perhaps this is true, and wise, but in an ordinary sense Wittgenstein was completely wrong. Death is *the* event in life. It is our chief organizing principle. It's why we rush and why we dawdle, why we butter up our bosses and fawn over our children, why we like both fast cars and fading flowers, why we write poetry, why sex thrills us. It's why we wonder why we are here.

Death comes most squarely into our lives at the places where we leave those who have died. In this respect, there is a noteworthy difference between a graveyard and a cemetery. A cemetery is a universe unto itself, an ocean of memories, each of which is always inexorably being carried further out to sea. A cemetery's vastness restricts the activity that takes place in it: the only reason to go there is to bury loved ones or pay respects. Graveyards, however, usually attached to old churches, are wedged into a human landscape, and everyday life has a tendency to wrap itself around them. Wander into an urban churchyard on a sunny day and you will probably find other people: kids playing tag, a homeless man sipping soup, people strolling, taking stock of their lives. This mingling in of ordinariness seems an unspoken nod to mortality, an acceptance—partial, anyway—of the fact that we, too, will actually die.

The Adolf Fredriks churchyard, in north-central Stockholm, is today an urban sanctuary rimmed by office buildings and shops. Tombstones scattered across the grass run a gamut of eras and funereal styles. There are tilting, centuries-old mini-obelisks, almost druidic in the angular cut of their tops, their faces weather-blasted of all records of the life that once occupied the remains beneath. There are art deco pink marble slabs, and rectilinear blocks with spare, geometric 1950s simplicity, neatly etched with crosses and the sans-serifed names of the departed: Johansson, Baggström, Thordal, Köpman. The yard is quietly busy; office workers come with lunches to sit among the graves. The most visited site is a raw, smooth, twisting monolith whose only marking is the swirl of a signature. In 1986, on the street that runs along the east side of the churchyard, Olof Palme, Sweden's outspoken leftist prime minister, was shot to death. The unassuming character of his gravesite suits a country famed for its spare design sense, and it remains a place of continual, understated pilgrimage.

Three and a half centuries ago, this was a forlorn little graveyard in the countryside, the out-of-the-way place Pierre Chanut chose for his illustrious friend's final rest. Christina ordered an imposing monument and had its four sides covered with Latin inscriptions, written by Chanut, extolling the epochal wisdom of the deceased and containing the names of both herself and Chanut.

The tomb is gone today, and so are the remains. In the spring of the year 1666—the first of May, in fact—with a strengthening sun warming the top layer of soil, coaxing life out of the dead land, a shovel bit into the precise spot of earth where, sixteen years earlier, a ceremony of supposedly permanent interment had taken place. Much had changed in those sixteen years. Most convulsively for Sweden, Christina was gone. Her enthusiasm for Greek esoterica had been short-lived, but the gossip about her interest in Roman Catholicism had proven to be on the mark. In 1654, four years after Descartes' death, she abdicated her throne, converted to Catholicism (in the staunchly Lutheran nation that Sweden had become since the Reformation it was insupportable that the monarch be a Catholic), and moved to Rome. There, as the most famous woman in the world and now either the most notorious or the most revered, depending on one's religious perspective, she had created an altogether new persona.

Christina's dramatic transformation—from enlightened monarch to religious convert of dubious sincerity (in Rome she continually flouted Catholic observances)—sparked incredulous speculation as the events unfolded, and the speculation has never ceased. Almost immediately people blamed—or credited—Descartes, whose commitment to his faith was well known, despite the charges of atheism that dogged him. But the contact between the queen and the philosopher had been limited and strained, so that, despite indications from Christina herself that Descartes had had a hand in her conversion, her biographers have looked elsewhere for sources of influence. Most have found it within her own nature: a deep, quixotic restlessness, a hungry, almost angry search for answers, for certainty. This, perhaps, was where she and Descartes had truly intersected.

Chanut, too, was gone. He had returned to Paris the year after Descartes' death and himself had died in 1662. The current French ambassador to Sweden, who watched as the shovel dug deeper and slowly revealed the coffin lid, was a very different sort of man. Where Chanut had been an enthusiastic promoter of science, a futurist who believed in the real-world possibilities of Cartesianism, Hugues de Terlon stood frankly with one foot in the past. He was a knight of St. John, a member of the chivalrous order based on Malta, whose glory dated to the First Crusade. Terlon was an imposing man of fifty-four, with a patrician nose, a thin, curling moustache, and eyes that had seen battle among northern European foes from Lübeck to Piotrków. He maintained an archaic and militaristic form of Catholicism. He was not only a warrior and a diplomat but also a member of a religious order that mandated a vow of celibacy.

Terlon was a loyal servant to Louis XIV, but he had been especially devoted to the Sun King's mother, Anne of Austria. Anne's husband, King Louis XIII, had died when their son was five years old, so that the queen took over the government as regent until he reached maturity. To aid her during this interregnum, she had deepened her already quite intense faith and had forged bonds with devout noblemen such as Terlon. Anne attended Mass several times a day and had a dizzying network of churches and convents at which she prayed on various feast days. She also took part in the revival of relic worship that had sprung up in the mid-seventeenth century. She owned several pieces of saints as well as a fragment of the Cross.

Because Terlon was also known for his piety, as he traveled

through the war-scarred landscapes of northern Europe on behalf of the French Crown, displaced members of religious orders sought him out for aid. Along the way, he, too, became a collector of holy relics. Friedrich Wilhelm, the elector of Brandenburg, one of the most powerful men in Europe, on meeting Terlon amid the chaos of battle outside Warsaw in 1656, gave him a coffer of relics that, he said, had been looted from a church in Vilna. In 1657, an advance guard of Swedish soldiers pillaged a convent outside Strasbourg. When Terlon arrived, the nuns emerged from the smoke and ruin and pressed him to take the convent's most sacred possessions, their relics, for safekeeping, which he did. In both cases, Terlon carried them with him on his journeys and, on returning to Paris, presented them to Anne, the mother of the king.

Now, in Sweden, Terlon had been asked to deal with relics, and he went about it with the same pious zeal. The unearthing of Descartes' bones that day in May 1666 must have run like a reversed film of the burial, for after the coffin was undug it was loaded onto a cart and hauled back down the same road, across the same bridge, and carried into the same building in which Descartes had lived and died. This was still the residence of the French ambassador, and Terlon was determined to keep the remains close at hand.

The sequence of events that unfolded next is important to understanding what was really taking place—as is the level of detail with which the matter of the shipping of a box of bones was committed to written record.

Terlon had been in the process of leaving Sweden—he was being transferred to the post of ambassador to Denmark—when he had received an official communication from the French government ordering him to quietly approach Swedish officials about the possibility of removing the remains. He asked for and received the permission, and now he took the rather extravagant step of engaging a contingent of Swedish soldiers to guard them around the clock as they lay in the chapel of his residence. Whether Terlon noted it or not, the captain of the Swedish guard, a man named Isaak Planström, seemed to take a particular interest in the assignment.

Terlon arranged for the body to travel with him as far as Copenhagen. He had a special copper coffin made that was only two and a half feet long. The improbable reason for this—besides the fact that the original wooden box in which the body had been laid had rotted—was that his superiors in France were concerned that if it became known that he was transporting the remains of René Descartes his party might be attacked and robbed. The cult of Cartesianism had grown strong in the years since Descartes' death, and others had developed an interest in the remains.

The manner of Terlon's subterfuge followed from the state of the remains after sixteen years. When he opened the coffin into which Chanut had had the corpse laid, he found that putrefaction was complete: soft tissue had gone, leaving bones that had loosened apart. A small box—just long enough to accommodate the largest leg bones—would be more inconspicuous than a coffin.

Another religious ceremony then took place in the chapel of Terlon's residence. The ambassador had assembled his embassy staff, other members of the French community in Stockholm, and Catholic priests—in fact, "nearly the whole Catholic Church of Sweden," according to Descartes' seventeenth-century biographer, who had access to Terlon's report of the event. That may not have been all that vast an assembly given the semipersecuted state of Catholicism in Sweden at the time, but it was clearly church sanctioned. They were all there to witness a ceremony of repackaging: to the accompaniment of formal prayers, the bones were taken from the rotted wooden coffin and put into the small copper box, where they were stacked one atop another in a manner that was deemed to be "without indecence."

Here Terlon paused the proceeding in order to make a request. He asked the assembled Catholic clergy if he might "religiously" be allowed to take one of the bones for himself. In particular, he had his eye on the right index finger—a bone "which had served as an instrument in the immortal writings of the deceased."

It's worth pausing to consider this request. Those in Paris who had worked through channels in the French government to have the bones removed and brought to France had their own reasons for doing so, which had to do with philosophy and politics, as will become clear in the coming pages. Terlon's interest was quite different but just as noteworthy. He was no mere courier: he was a knowledgeable man of the world. Descartes' fame had spread in sixteen years, and both Cartesianism and the man whose name it bore were the subject of rumors, hopes, and fear. Here in Sweden, the estate of the clergy, the branch of the government controlled by the Lutheran Church, had tried, two years earlier, to outlaw the teaching of Cartesianism, so great was the threat they felt from it. In Leiden and Utrecht, after the first debates about Cartesianism had flared up, the philosophy had become rooted, so that while Terlon was overseeing the disinterral Cartesianism was growing in strength in nearly all branches of the universities. In France, Spain, Germany, and Italy, men who had been students during Descartes' lifetime were now professors, churchmen, and physicians and had brought with them certain convictions about the truth of the Cartesian method of acquiring knowledge and the overall approach to nature that it implied. In each place, a battle was intensifying and becoming more complicated. It wasn't a case of "science" versus "religion," or even of the purely new versus the utterly old. Some, like Gysbert Voetius at Utrecht, believed the materialism in the new philosophy was a direct attack on Christianity. At the same time, however, important Jesuits and Oratorians—two of the most prominent and intellectually driven Catholic orders—had become Cartesians and actually saw the philosophy as a way to protect the faith.

Then there were those like Terlon for whom any inquiry into the heart of nature was at the deepest level a spiritual inquiry. What is the nature of light? Why does salt form crystals? How does the experience of fire touching the skin transmit through the body and record as pain in the mind? We think of such questions as frankly the realm of science, but for a seventeenth-century European, nature, including the human body, was inarguably the terrain of the Almighty; to come to understand it more fully than had ever been done was, as it were, to touch the face of God.

Today we associate the reverence of relics mostly with the Middle Ages, but while the Council of Trent, a century before Descartes' death, put an end to the commercial trade in relics, Catholic theologians continued to stress their importance, and ordinary people as well as the highborn continued to venerate them. The veneration of relics meant more than the mere honoring of a great person but less than worship. It was a deep meditation on the physical being of humans and on the body as a "temple of the Holy Ghost." We are used to thinking of Christianity's emphasis on the afterlife and its view of the body as sinful, but Catholic tradition in the early modern period emphasized the physical. Bodily remains were keys to the deepest of mysteries, links in the chain between life and death, and, as the Council of Trent said, the bones of prophets, saints, and others "now living with Christ . . . are to be venerated by the faithful." In asking for a relic of Descartes, the chevalier de Terlon was standing at the cross-roads of the ancient and modern. He was applying to a modern thinker—the inventor of analytic geometry, no less—a primitive tradition that extends back not only to the institutionalization of Christianity in the fourth century, when Christians first broke into the tombs of saints to gather relics, but farther still, beyond the horizon of recorded history. The request is all the stranger for the fact that the man whose remains were treated in this quasi-saintlike way would go down in history as the progenitor of materialism, rationalism, and a whole tradition that looked on such veneration as nonsense.

The priests granted the request; the chevalier was allowed to take the finger. He must have kept it until his death in 1690—he certainly didn't give it to Anne of Austria, who had died earlier that year—perhaps keeping it on his person as he traveled the next ten years between Paris and Copenhagen. On his death he was required to bequeath his property to his order, the Knights of St. John. The inventory of the order contains no artifacts from Terlon and no index finger labeled as Descartes'. Terlon's branch of the order, based in Toulouse, was, like many other Catholic holy places, pillaged during the French Revolution. Perhaps the finger bone of Descartes—we might call it the first modern relic—slipped through the fingers of a sans culotte, into the dirt and out of history.

The verb *to translate* has a particular meaning in a Catholic context. In the year 787, the Second Council of Nicaea ruled that the new churches that were proliferating across Europe should each be anchored in sanctity by a holy relic. This ruling created an official market in bones, as priests and bishops sought portions of prominent or relevant saints for their new churches. The transfer of relics from, say, a tomb in Sicily to a church in Lombardy was referred to as a translation, and throughout the Middle Ages and into the early modern period holy bones in translation—housed in boxes of precious metal, adorned with drapes and candles—were part of the traffic on the highways of Europe.

Terlon's translation party left Stockholm in June 1666. The remains were loaded onto a ship and placed under the care of two members of Terlon's staff, the sieur de l'Epine and the sieur du Rocher. There was a fuss at the port when the sailors learned that human remains were part of the cargo, which in their lore spelled bad luck, but Terlon managed to quell the uproar, perhaps by showing them the small copper box and convincing them that they weren't so much shipping a dead body as translating relics.

Terlon was anxious about theft. Descartes' seventeenth-century biographer, Adrien Baillet, wrote that the fear was that "this precious cargo would fall into the hands of the English, among whom Descartes had an infinity of worshippers . . . and who would build a magnificent mausoleum in their country, under the pretext of erecting a temple to Philosophy." Before the ship left, Terlon wrote to Louis XIV informing the king of the steps he had taken with regard to Descartes' bones and reminding His Majesty of the illustriousness of the deceased. Louis wrote back granting royal authority for the translation of the remains. Not content with this, Terlon also personally disguised the copper reliquary, giving it "the appearance of a bundle of rocks." The ship then set out from Stockholm bound for its first port of call, Copenhagen.

It's unlikely that any pilgrim has ever retraced the translation route of this particular collection of relics. From Copenhagen, the party, which from here would be led by the two members of Terlon's staff, set out on a morning in early October, heading south. They made an uneventful passage through the wilds of northern Europe—Jutland bogs, North Sea coastal marshes, fog-

bound villages of Lower Saxony, through heath and forest and ultimately into the flat expanse of Flanders—until they reached the
northern French town of Péronne. Here, customs officials took
an interest in the train; finding the curious package and discovering that its outward appearance belied a bright copper interior,
they demanded that the knights open it, making it clear that they
suspected contraband. L'Epine and du Rocher affected official indignation; they produced a letter that Terlon had given them from
no less an official than Pierre d'Alibert, the treasurer general of
France; they pointed to the ambassadorial seal that Terlon had affixed to the box. But the officials wanted it opened, the strong iron
bands that Terlon had taken the precaution of wrapping around
the box perhaps increasing their suspicion. In the presence of wit-
nesses, the bands were snapped, the box was opened, the officials
peered inside . . . and it was as they had been told, maybe even
less noteworthy, for, due to the rotting of the original coffin, parts
of the skeleton were reduced to fragments, on which individual
bones now rested. Presumably, however, they didn't sift or perform even a perfunctory inventory, because there was a striking
observation to be made, and nobody made it. A vital bone—the
most obvious part of a skeleton—was missing.

The box was resealed, horses were retethered, and off they went
again, headed, without further interruption, for Paris.

E VERY WEDNESDAY EVENING IN the late 1650s and
through most of the 1660s, a cross-section of French society could be found packed into a house on a narrow alley in Paris
known as the rue Quincampoix, a few steps from the raucous and
reeking market of Les Halles. The mix was atypical for the time,

almost scandal-worthy. Women and men, both single and married, were thrown together, high government officials alongside uncouth provincials, as well as princes and prostitutes and canons of the church—a profusion of frilled collars, puffed sleeves, and flowing, curled hair filling all three floors of the home. Today the narrow building lies just a few steps from the pedestrian-only rue Rambuteau, lined with kebab sellers, piercing booths, and shops selling paté sandwiches and the ubiquitous Robert Doisneau photos of Paris in the 1950s. The upper floors of the house remain residential; the street level is a karaoke bar. Three and a half centuries ago, as the home of Jacques and Geneviève Rohault, the building was well furnished, ornamented with tapestries and paintings, but, more remarkably, strategically littered with beakers, tubes, syringes, microscopes, prisms, compasses, magnets, and lenses of various shapes and sizes, as well as such curiosities as an "artificial eye" and a large mirror affixed to the floor.

The visitors to *les mercredis,* as these weekly happenings became known, included some of the most famous names of the century, among them France's supreme playwright, Molière, the socialite Madame de Sévigné, and the Dutch polymath Christiaan Huygens, who invented the pendulum clock, discovered celestial bodies, and helped develop calculus. They all came to see Rohault, a physicist who was known as the greatest living Cartesian. This description in itself suggests the gulf between that time and ours. Today, or twenty years ago, or a century ago, a student of philosophy learned about Descartes as a philosopher, a man who refashioned the landscape of the mind. For the generation that succeeded him, Descartes was that, but he was also an investigator of nature, and, for these men and women, the two things were intricately connected.

In 1667, Rohault was forty-seven years old, built like a bulldog, with a personality to match. He could be short to the point of irascible with those who failed to grasp a principle he had repeatedly tried to explain. At least part of the shortness came as a by-product of zeal. His devotion to the master was such that he had married into the circle of Cartesians. His wife was the daughter of Claude Clerselier, caretaker of Descartes' writings (many of which were as yet unpublished). Like Descartes himself, Rohault was intoxicated with the new means of comprehending the physical world and seemed to believe that any observation, any datum, from whatever field of inquiry, was liable to tip the balance from ignorance toward knowledge and mastery. He had made himself an authority on astronomy, geography, and anatomy. He wrote a detailed commentary on Euclid's geometry and how to put its principles to work; his *Traité de physique* remained a standard text-book on physics for decades. And he didn't restrict himself to the scholarly realm but roamed the alleys of Paris in which various craftsmen practiced their trades, watching them build clocks and distill brandy, querying them, trolling for notions and clues and methods.

This combination of intellect, acute observation, and missionary fervor came together in Rohault's weekly demonstrations. There was an element of performance involved, and no doubt some of those who came did so for the show. Colored flames, bubbles, explosions: Cartesianism had become a spectacle. It was also rumored to offer glimpses beyond the material, into the realm of the supernatural, and some were titillated by the possibility of peeking behind the screen of ordinary existence. But Rohault and most of his visitors were after something else. There was an order in nature; one could proceed from the unseen bedrock of philosophical principles—the Cartesian method—by rungs up into the realm of matter and its manipulation. His weekly event was a salon, but one in which people took notes, scribbling fast as they tried to grasp something genuinely new.

The notes taken by an anonymous lawyer who frequented the *mercredis* have survived. This *avocat*'s account of one evening gives a sense of the scope and the intensity that people from all walks of life brought to the abstract stuff of philosophy. Rohault began the lecture by discussing the two states of waking and sleep—"The first sees true," the lawyer wrote, "the second false"—followed by a discourse on the study of dreams. In these introductory words, Rohault was asking his audience to develop perspective on the mind and its proper functioning. Then he moved into an analysis of Descartes' reorientation of how we know. The key, he said, was the "cogito," which gives us certainty about our individual existence. Everything else, the lawyer noted, is only "probability." And there, in his carefully wrapping that strange and deeply modern word in quotation marks, this nameless amateur philosopher gives us a shiver of what it must have been like, sitting amid the Louis XIV coiffures and the chalk-and-vinegar aroma of face powder, to first glimpse the horizon of the modern world.

From here, Rohault moved into Descartes' division of reality into mind and matter. Remarkable—maybe even limitless—improvements were possible in every field of endeavor, but faster coaches, stronger swords, and more sensitive lenses required a truer understanding of the physical world and how we know it.

What exactly constitutes a material object? Attempting to answer that question plunged the physicist and his audience into the depths of philosophy. "In connection with the 'belief' in the existence of the 'physical being,'" the lawyer noted, "we seek to understand what it is that persuades us of this belief. For example, heat is not the essence of a thing, because there are cold things. Nor is coldness the essence, because there are hot things." And:

"It is neither the hardness nor the liquidity which is the essence of a thing, and for the same reason."

Here Descartes and his followers had set themselves in opposition to the Scholastic notion of matter. According to the traditional way of thinking about these things, the sky has blueness in it, water has wetness, garlic has its odor. These perceivable qualities are embedded in the underlying substance of a thing. Rohault, in particular, found this to be faulty logic that impeded material progress. The Aristotelians, he wrote, reason that "it would be impossible for luminous or colored bodies to cause those sensations in us which we feel, if there were not in them something very like what they cause us to feel, for, they say, nothing can give what it has not." Rohault dismantles this logic with a simple example: the fact that a needle poked into the skin causes pain clearly does not mean that the pain is somehow in the needle. The pain is in the mind. And so, in some sense, is the blueness of the sky, the wetness of the water, the smell of the garlic. This is the root of what Descartes bequeathed not only to his immediate disciples but to most of the rest of us who came after. There is a divide in the universe—there are two distinct substances. One is matter. The other is mind. Reality is not "that out there" but a dance involving the sensor and the sensed.

Abstract as this all is, the notion was of the essence for the Cartesians. Strange to say, it was also dangerous. These conceptual, seemingly otherworldly notions had political import. As much as the Cartesians themselves wished it were not so, Cartesianism threatened certain centers of worldly power. As they met at Rohault's *mercredis* and other, sometimes fancifully named Cartesian salons in Paris and around Europe (the Société des lanternistes in Toulouse, for example, took its name from the torches its members carried to light their way to its evening sessions), these men and women were acutely aware of doing so under a threat, which eased and intensified as various church and state officials modified their understanding of the new philosophy. In the case of each institution, the fear was of having its power undermined. If this new sect, Cartesianism, professed to be able to show, for example, that the body was a kind of machine and death was an absolute barrier, then where did that leave the doctrine of the afterlife or of Christ's bodily resurrection? If miracles could be authoritatively dismissed as nonsense, a faith built upon the miraculous was groundless. For an autocratic government the threat was equally serious. While Rohault was giving physics demonstrations in Paris, in Amsterdam Baruch Spinoza—also picking up where Descartes left off—was using reason as the base from which to argue that democracy, not absolute monarchy, was the only just form of government. Such ideas were whispered in Cartesian circles, leaving rulers around Europe to view those circles as suspect, if not treasonous.

All of these fears became concentrated in a single issue. It wove one of the most abstract and seemingly unworldly elements of Cartesian philosophy into individual human life, society, and worldly power. This issue was perhaps the biggest source of anxiety that Descartes himself had for his philosophy. As for the Cartesians who sat in their salons watching demonstrations of the "magic lantern" (a forerunner of the slide projector and the cinema) and witnessing experiments involving mercury, magnetism, and barometric pressure, it gave them palpable fear of a very real-world kind: of soldiers appearing at the door to lead them away. The issue concerned the truth of the Catholic sacrament of Communion.

A notion had first struck Descartes in 1630, in a seemingly innocuous way, when he was thinking about optics and color. When you break open a loaf of bread, the inside is so very white. Surely that whiteness is in the bread itself, is it not? From this mundane hook his mind wove a chain of logic that threatened the major institutions of Europe. Descartes himself would rather not have explored it further, but in 1643 he received a letter from one of the teachers at his old college of La Flèche. Père Denis Mesland had become a devotee of Descartes', and he now had some questions. In the seventeenth century, as now, the central rite of Catholicism was the Mass, and the center of the Mass—the essence of the faith—was Holy Communion, in which celebrants receive bread and/or wine: the "body and blood" of Jesus Christ. One of the chief differences between Catholicism and Protestantism—one of the spurs behind the century of bloodshed that was just then ending—involved the meaning of the quotation marks in the previous sentence. Protestants (some of them, anyway) came to hold that the bread and wine *represented* Christ's body and blood, whereas for Catholics mere symbolism did not get at the genuine nature of the sacred mystery involved. In Catholic theology (and in Catholic conviction, in the seventeenth century and now), when a priest repeats during the Mass the words Jesus spoke at the Last Supper—"This is my body . . . this is the cup of my blood"—he initiates a real conversion of substance. A century before Mesland's letter, the Council of Trent, which formed in reaction to the Protestant Reformation, had decreed that, regarding the consecrated bread and wine, "our Lord Jesus Christ, true God and man, is truly, really, and substantially contained under the species of those sensible things."

For the church, Christ was, and had to be, "really, truly, and substantially" present in the bread and wine. Intelligent, reasoning Catholics need not necessarily have a problem with the logic in this, thanks to their belief in mystery as a real force in the world.

Indeed, the transformation could not be explained by ordinary means; this was part of the essence of the faith, just as the bodily resurrection and ascension to heaven were elements of the mysterious truth of Jesus' sojourn. As for how it was that the bread, after the priest had performed the ritual of consecration, still looked like bread and felt like bread and tasted like bread, Catholic theologians had worked it out using Aristotelian categories as adapted by Thomas Aquinas. A material object, in Aristotelian science, is comprised of accidents—color, odor, taste—and substance, the real underlying thing itself. When a priest blesses bread and wine during the Mass and repeats the biblical formula, the transformation happens at the level of substance. The underlying *substances* of the bread and wine are swapped for the substances of the flesh and blood of Jesus Christ. Thus the term *transubstantiation,* which came into use by theologians around AD 1100. But the *accidents* of the bread and wine—which give them their appearance but which are actual components of the bread and wine—are left unchanged. This seemingly problematic bit of reality was in fact held to be a second miracle, and in the Middle Ages "proof" of the twin miracles of transubstantiation occasionally manifested itself in the world. The American philosopher Richard Watson, in writing about the Aristotelian explanation, likened the notion of accidents miraculously keeping up a faux appearance to a "shield" that covers the real substance and added, "Numerous stories were known of the shield having been dropped, so that the priest saw lying in his hand an actual piece of flesh or, more spectacularly, a tiny, perfectly formed baby."

As far as Catholic authorities were concerned, the physics of transubstantiation had to be explained in this way. It is difficult to overestimate the multifaceted importance of the Eucharist (to give the sacrament of Communion its proper name) in Catholic Europe in the seventeenth century. Because the host actually becomes Christ's body, it has incorporated into it Christ's bodily pain, his willingness to suffer and die for humanity, and thus his love for humanity. Consuming the host is an act of recommitment to the faith, participation in his suffering, and acceptance of his divine love. To eat the bread is also to unite one's own physical body with Christ's—to become part of the body of Christ, meaning both his physical body and the body of believers. In this way, the ritual of the Eucharist was, and remains, the very essence of Catholic Christianity, tying the esoterica of faith to the essential fact of humanity: our physical, flesh-and-blood selves. Again, the point is that the act isn't symbolic but real. The Catholic host brings what Christians believe to be the historical root of the faith—that, in Jesus, God became flesh and suffered physically and died—into the here and now, over and over, every time the Mass is celebrated.

As spiritually meaningful as this concept was, it also had tremendous implications for real-world power. The whole infrastructure of the church—parishes and cathedrals, priests and nuns, real estate, art, revenue, the ability to mold and manipulate heads of state—rested on it. Only an ordained priest could say Mass, and in celebrating Communion, repeating the words "This is my body," the priest took on the personage of Christ and became the indispensable vehicle through which Catholics participated in the mystery of Christ's bodily suffering and death and resurrection. Because the host was the real substance of the body of Christ, the church had what it believed to be the franchise on salvation. The Protestant Reformation represented an assault on transubstantiation and the real-world power it gave the Catholic Church. Father Mesland was among the first to recognize that Cartesian science was another such assault. In Descartes' reckoning of the universe there was no "real" apple or tree or butterfly lurking beneath the shield of accidental appearances of these entities. If the object in question was hard, gray, and flecked and otherwise gave every appearance of being a piece of granite, then it *was* a piece of granite. And if it looked, smelled, and tasted like bread, it was . . . bread. That was the direction Mesland saw Descartes' ideas heading, and it was a dangerous one.

Descartes responded with an assurance that his philosophy did not deny the genuine presence of Christ in the host. In fact, he believed he offered a philosophically satisfactory account of it, one that could coexist with the mechanistic view of nature. Indeed, by stressing a dualistic view of reality, by putting the ephemeral stuff of mind and soul in one category and the physical world in another, he believed he was building a wall around the fortress of faith, protecting it from the encroachments of science. At the same time, he was hoping to protect investigations of the natural world from theological interference. He had been shaken by church condemnations of scientists, particularly of Galileo ("I was so surprised by this that I nearly decided to burn all my papers, or at least let no one see them," he wrote on learning of Galileo's conviction for publishing his heliocentric views). Descartes himself was so devout in his faith yet so certain of the legitimacy of reason-based investigations of the natural world that the division of reality into two distinct halves seemed the only logical conclusion.

While the goal was in part to protect religion, one long-term effect of Cartesian dualism—as it seeped into Western consciousness over the ensuing decades and centuries—was to drastically limit religion's scope. In the prevailing modern view, faith has no business meddling in astronomy or biology. And the logical extension of this thinking has been the very modern stance of atheism. To some extent, such an outcome was foreseen by critics of Descartes' own time, who saw his work and that of other mechanistically inclined philosophers as giving reason full sway over human reality and relegating faith to superstition. This was surely not Descartes' intention, nor was it the intention of his contemporaries. Descartes' lifelong timidity in confronting church authorities was always at odds with his ambition, and in the matter of the Eucharist he hoped to have it both ways: to lay a new foundation not only for physics but for Christian theology. He pushed Mesland, and other of his followers, to take up the challenge and bring the church's explanation of transubstantiation in line with science—that is to say, Cartesianism. Again, his immodest goal was to replace Aristotle completely as the base of all knowledge.

Descartes himself avoided direct attack over the Eucharist, but from the time of his death the matter expanded into a full-fledged controversy. His followers took up the challenge in various ways. Descartes had tried to override the whole mechanism of the Aristotelian explanation, arguing that it was a mistake to talk about transformation of substance, that instead the miracle involved the union of Christ's *soul* with the bread. In this way, there was no need for a second miracle, in which the "shield" of breadlike appearance covers the underlying substance. This explanation itself caused alarm, since it seemed only a slight variation on Protestant ideas that the host symbolized Christ's body. For the church, the soul of Christ was apparently not substantial enough to support its worldly edifice; Catholic authorities needed the body, too.

Nevertheless, the Cartesians pushed their arguments. Rohault offered a defense of the Cartesian view of the Eucharist. Claude Clerselier, Rohault's father-in-law, wisely refrained from including the exchange of letters between Descartes and Mesland in his publication of Descartes' correspondence, but he sent copies to influential parties. Robert Desgabets, a Benedictine monk with a penchant for science who had never met Descartes but had become entranced by his philosophy, was one of the recipients of the letters. Desgabets journeyed to Paris to join Cartesian salons, and—rather dramatically demonstrating the close link the Cartesians saw between philosophy and medicine—lectured on how one might perform a blood transfusion while also offering his own support for a Cartesian view of transubstantiation.

After Desgabets left Paris, he toured Benedictine abbeys in the countryside to spread the gospel of Cartesianism. Desgabets eventually published a text whose title spelled out the central matter pretty clearly: *Considerations on the present state of the controversy touching the Very Holy Sacrament of the host, in which is treated in a few words the opinion which teaches that the matter of the bread is changed into that of the body of Jesus Christ by its substantial union to his soul and to his divine person.* Together with his other activities, this little text got Desgabets branded as a heretic, with the result that his work was suppressed and his name largely forgotten by history. Meanwhile, Father Mesland, for his persistence in pursuing the question, was eventually banished to Canada.

This then was the climate of increasing danger in which the Cartesians operated. Still, some—Rohault among them—continued to argue that their principles could be put in the service of both the ruling civil and spiritual authorities. Far from being a threat, they claimed, the new philosophy could be the protector of the faith. Descartes himself had taken this line. There were many in power who were intrigued by the idea of this strange and dimly understood new tool actually becoming part of the arsenal of the church or state. The climate alternated between curiosity and fear. The situation of the Cartesians in the late seventeenth century thus mirrors in some way that of the early Christians in the catacombs of ancient Rome. They were alternately tolerated, suspected, then persecuted—and of course eventually triumphant in the spread of their philosophy.

There were other parallels that existed in the seventeenth century between Descartes and Jesus. Many of the early Cartesians were themselves Catholic priests. In some sense the new philosophy was to be a replacement for Christianity as the foundation of Western culture, and indeed the Cartesians referred to themselves as "disciples of Descartes." Their physics collided with Catholic views about the body of Christ, and they were about to use the material body of Descartes, or what remained of it, to promote their philosophy. Then there was the fact that during his life Descartes had seemed to believe that he could somehow override death's dominion—the irony being that his "eternal life" idea rested on scientific rather than religious beliefs.

Also like the early Christians, the Cartesians believed devoutly in their cause. Some held it in almost mystical regard. They were keepers of a legacy and carriers of a flame that they believed would light the future of the world. They knew that what they were about was dangerous, and that it required knowledge not only of the intricacies of philosophy and science but of how power worked. In order to survive and advance their cause they needed to employ tools of persuasion. And now, at the turning of the year 1667, a new tool was about to arrive.

I T WAS IN THE COLD of January, three months after they had set out, that the two Frenchmen, l'Epine and du Rocher, arrived at the outskirts of Paris. Long as the journey had been, as with any modern road trip, reaching the metropolis would have meant slowing down further. Paris was still largely a medieval city, with dirt roads that were only beginning to be paved and streets in an irregular tangle. It was bigger, noisier, and dirtier than London, and the large-scale improvements that Louis XIV's chief adviser, Jean-Baptiste Colbert, had recently begun—driving through wide boulevards, clearing away crumbling parts of the old medieval walls, constructing the colonnade of the Louvre—at this stage merely added to the congestion. People complained, but the king stayed mostly at his Versailles palace, visiting Paris only a few times a year and otherwise keeping it out of mind. The most noticable change from the last time Descartes had visited the city in life was visible in the streets. Early in the century, transportation was either on foot or on mule. By now a revolution had occurred: vehicles of every description, thousands of them, from simple rickshawlike contraptions pulled by men known as "baptized mules" to gilt carriages with glass windows and shock absorbers, clogged the streets, so that the wagon carrying Terlon's parcels, including the remains of the philosopher, would have had to weave a contorted passage. They would have come through one of the two crumbling medieval gates at the north end of the city, the Porte St.-Martin or the Porte St.-Denis, made their way across the fashionable neighborhood of Le Marais, and come to stop, finally, at a stately residence in the midst of bureaucratic Paris, just north of the Seine.

Soon after the wagon arrived at the home of Pierre d'Alibert, treasurer general of France and the highest-ranking Cartesian in the French government, Jacques Rohault, in his home just a few blocks to the west, Madame de Sévigné, who lived an even shorter distance away, and other Cartesians got word of it. Foremost among these was Claude Clerselier, the fifty-three-year-old government official who had been one of the earliest converts to

Cartesianism. During Descartes' life, and even more so after his death, Clerselier made himself indispensable as a kind of literary agent. Descartes had deemed some of his writings too incendiary to publish during his lifetime; Clerselier edited them and oversaw their posthumous publication. He also published the philosopher's correspondence, which wasn't easy. Descartes' effects, including his letters, were shipped to France after his death in Sweden, but the ship sank. Clerselier took charge of the salvage operation, employing a team of workers to recover the thousands of sheets and dry them out.

Through the 1660s and 1670s, Clerselier kept the volumes of Descartes' works appearing in print, giving the educated classes of late-seventeenth-century Europe fresh intellectual fuel and both keeping Descartes himself in living memory and using his works to further the cause. In doing so, he became the leader of the Cartesians—not in a philosophical or scientific sense but as a general or chief strategist. The followers of Descartes formed a diffuse group that ranged all across the Continent, but its core was remarkably tight knit. Those in the inner circle were linked not only by their devotion but also by ties of blood and marriage. Rohault was married to Clerselier's daughter, while Clerselier had married Pierre Chanut's sister.

It may have been Clerselier—who had been working for so long on the body of Descartes' writings—who came up with the idea of putting the physical body of the master to use as well. Sixteen years earlier, Chanut and Christina had decided, for their own reasons, to bury the philosopher in Stockholm, but soon after Chanut's letters announcing the death reached France certain parties began to clamor for a translation to French soil. Nationalism may have been the initial motivation, but in time another thought dawned, and the irony in it probably did not escape their notice. Descartes' revolutionary philosophy had been rooted in his focus on bodily health, in particular on his own body; that philosophy had recently come into conflict with official views about the physical body of Jesus Christ. Now, in their effort to legitimize that philosophy, and to protect themselves, Descartes' followers would put his own physical remains into service.

It took months to organize the sanctification of Descartes. The Cartesians—Rohault, Clerselier, d'Alibert and others—laid out their plans with the exquisite orchestration of political operatives. Their goal was to influence people in church and government, so the effect they hoped to create was one of power and force, something inexorable, that demanded official sanction and respect. Finally, on an evening in late June, their spectacle was ready. As the sun slowly set, a vast group of people gathered in the narrow street in front of d'Alibert's home just off the Seine. Clergymen, aristocrats, and friends of the philosopher were among the crowd. But, just as important, large numbers of ordinary Parisians, including some of the city's poorest inhabitants, filled the street. The poor were given flaming torches to carry; the rich rode in carriages. They formed a procession, heading north to the rue St.-Antoine. Here they turned left and came to the blocky edifice of the church of St. Paul, in whose chilly interior the coffin containing the remains of Descartes had been left since the winter. Bearing it with them, the funeral cortege headed south, making a dramatic bisection of the city, crossing the Ile de la Cité to the Latin Quarter, and following the rise of the land until they reached a broad, breezy summit. As they arrived, church bells rang out. Two churches stood side by side on the hilltop. Before the one on the right—the Gothic Ste.-Geneviève-du Mont, named for the patron saint of Paris—stood its abbot, the Reverend François Blanchard, in full habit, a miter

on his head and a crucifix in his hands. Next to him were the
canons of the church, each holding a lighted candle.

Into the church the whole procession filed. D'Alibert had given
the abbot suggestions as to the type of religious spectacle he might
mount, but the abbot exceeded them. The procession itself refer-
enced the annual pilgrimage from the church through the streets
of Paris carrying the bones of St. Geneviève. The spectacle aroused
the city—the next day, even larger masses of people gathered in
the vast open square before the church to see what was going on.
Clerselier had planned for a public funeral oration, but at the last
minute an order came from Louis XIV's government forbidding
it, so it was given in private. The prayers and celebration culmi-
nated with the coffin's being brought to a vault of honor that had
been designated for it, next to the remains of St. Geneviève.

After the abbot blessed the remains of Descartes, the proceed-
ings moved into an administrative phase, in which a series of formal reports was offered. The Cartesians presented to the church
a written account documenting the steps in "this famous translation," including statements by Clerselier and the late Chanut,
certificates by religious authorities testifying to Descartes' un-
swerving Catholic faith and the "exemplary innocence" of his life,
and a remarkable letter written from Rome by Christina, in which
(still maintaining the royal pronoun) she certified that Descartes
had "contributed greatly to our glorious conversion." All of this
data, so carefully compiled, was meant to dispel the notion that
Descartes himself—and by extension his philosophy—was in any
way anti-Catholic or, for that matter antireligion. Finally, Clerselier presented a copper sword on which the details of the whole
translation from Sweden and the ceremony in Paris were engraved, along with the names of the most prominent men present.
The abbot placed the sword in the coffin "in the presence of these

friends," as Descartes' seventeenth-century biographer noted (a detail that will become important later), the vault was sealed, and the ceremony ended.

Then the parties began. D'Alibert hosted the main banquet, which included as many important people as Clerselier and his cohorts could muster; other feasts—"splendid and sumptuous"— were held around the city. Dukes, lawyers, mathematicians, courtiers, members of the Parlement of Paris and of the French Academy, the king's supervisor of fortifications as well as his physician all took part in the extended banquet over the returned bones of the philosopher. There was one purpose here: to advance the cause, to demolish, once and for all, the tenacious old structure of knowledge and win society over to Cartesianism, to the cult of reason, to the belief that the truest and trustiest foundation was the human mind and its "good sense." At the same time, the model—Catholic treatment of holy bones as relics—was so closely copied in all its particulars that it isn't even right to speak of the reburial as a secular co-opting of a religious event. It *was* a religious event—an attempt to carry the scientific perspective into a world circumscribed by religious awareness.

In part, the program succeeded. As one measure of success, over the next few years Descartes himself, in various guises, became a fixture in society. Stories and poems featured him. He became, as the French historian Stéphane Van Damme puts it, "a fictional literary personage who inhabited literary writings," appearing in some works as a man of science, in others as a disembodied thinker, and in still others as a spiritual presence. Literary games were a fashionable pastime in salons, and in the 1670s Descartes entered these as a character. One variety was a type of séance; in a poem referring to one such game, a young woman holds a conversation with Descartes' "illustrious and learned ghost."

In the years after the funeral over the translated remains, then, Descartes achieved a different type of popularity, and as he gained force as a cultural character his philosophy rooted in new places. Important aristocrats and churchmen—the prince de Condé, the duc de Liancourt, Cardinal de Retz—took Cartesians under their protection. As the Cartesians accepted these powerful patrons, Cartesianism became adopted by what might be described as opposition parties in both church and state. The prince de Condé, one of the most famous and dashing figures in Europe, had rebelled against Louis XIV and become leader of the opposition during the Fronde, the French civil war, which had ravaged the country between 1648 and 1653. He was now technically loyal once again, but from his castle in Chantilly he led a kind of rival power structure to that of the king, one element of which was the circle of Cartesians. Cardinal de Retz had also opposed the king in various ways; in punishment Louis had recently forced him to give up his position as archbishop of Paris, but he stayed active in politics and remained a champion of Cartesianism. Even the site chosen for the repose of Descartes' remains had "opposition party" associations. As revered as the church of Ste.-Geneviève was, it and its abbot had a long history of struggles with both the king and the diocese.

The establishment hardened and clarified its stance. In 1671, Louis XIV came down decisively against the Cartesians. The new philosophy had penetrated the University of Paris, where it was infecting the young, causing them to question beliefs about the Eucharist, the Mass, and the authority of the priesthood. The archbishop of Paris asked Louis to act, and he did, forbidding teaching that "could bring confusion in the explanation of our mysteries." Later he clarified that his opposition was to "the opinions and sentiments of des Carthes." The city of Paris followed suit by decreeing that anyone promulgating such doctrines would face the death penalty. One ironic result of the state condemnation was that when the French Academy of Sciences was formed in 1666, in an effort to bring the disparate salons and gatherings of freelance philosophers under state control, the Cartesians—men like Rohault, people who were doing the closest thing to scientific work—were expressly forbidden from joining.

Official Catholic opposition began in earnest even before the translation of the remains. In 1663, the Holy Office of the Catholic Church—that is to say, the Inquisition—condemned four of Descartes' books and placed them on its index of banned works. Its reasoning was kept secret until 1998, when Cardinal Joseph Ratzinger, the future pope Benedict XVI, who then held the title of prefect of the Congregation for the Doctrine of the Faith—the modern variant of Grand Inquisitor—ordered its files open through the year 1903. The dossier on Descartes' censure reveals depths of interest on the part of those studying the new philosophy, one of the churchmen writing, "I admired the author's subtle spirit, his inventive new speculations, the elegance of a style far from ordinary, and the modesty with which he submits his work to the censure of the theologians, and I pay homage to the merits of this writer." The end result, however, was condemnation and forbidding of Catholics to read the books on pain of sin, and the files show that of crucial concern to the church was the conviction that Descartes' reckoning of matter and the material world undermined the doctrine of the Eucharist and the real presence of Jesus Christ in the host.

Yet the boundaries of the struggle over Cartesianism were constantly shifting as people of various backgrounds worked to understand just what it was and what it might mean. While the Vatican opposed it, many if not most of the Cartesians were themselves churchmen, and some important orders within the church adopted it—so that besides the battle between Cartesians and those who held what was still the establishment worldview, built around the Bible and Aristotle, there were also struggles among the Cartesians for control of the philosophy and what it represented. And if Descartes' body could be used in this struggle, so could his life. Church leaders who had "converted" to Cartesianism but wanted to keep it within a church context succeeded in getting the first comprehensive biography of Descartes written by one of their own, Father Adrien Baillet, who also wrote a seventeen-volume *Lives of the Saints* and who gave Descartes' life story a similarly saintly cast.

Most significant for history was the fight among the Cartesians over the meaning of the philosophy. Three years before the reburial of the bones, a priest named Nicolas Malebranche was browsing the bookstalls that lined the banks of the Seine (as indeed they still do) and came upon a copy of Descartes' *Treatise on Man*. It was his first encounter with the new philosophy, and it struck him so forcefully that as he read he experienced shortness of breath and heart palpitations. He became one of the dominant Cartesians and, in particular, an advocate of what might be called Cartesian theology. Cartesians were in agreement in believing that the mind was the base of knowledge. But do ideas have a deeper foundation than the famously shifting and untrustworthy mind? Malebranche claimed that Descartes was clear on this. To talk of Cartesian dualism is somewhat misleading; Descartes actually wrote that the universe consisted of not two but three substances: mind, body (that is, the material world), and God. God is the guarantor that the mind and world can interact meaningfully—that we can reach truth using the power of reason. Malebranche insisted that human intellect and its ability to make sense of the natural world rested on God. As he put it, "We see all things in God." Malebranche thus interpreted Descartes as being an apologist both for science and for faith. Other Cartesians—including Rohault and the Parisian priest Antoine Arnauld—read Descartes as asserting that ideas exist in the mind of their own accord, as it were. While they didn't dismiss the relevance of God, they also didn't feel that God necessarily played a role in the relationship between the world and the mind that perceives it. The difference between this and Malebranche's interpretation of Descartes may seem slight, but it would widen, ultimately, into the idea, so characteristic of the modern centuries, of a firm division between faith and reason and would lead, seemingly inexorably, to the modern concept of atheism.

These battles—between Cartesians and the church, between Cartesians and state power, and among the Cartesians themselves—raged through the dying days of the seventeenth century and into the eighteenth. There would be casualties among the Cartesians. People would be excommunicated and forced into exile. Rohault himself would be accused of heresy and banned from holding public lectures.

More importantly, other thinkers—the German Gottfried Leibniz, the Amsterdam Jew Baruch Spinoza—would spin Cartesianism in new directions, broadening the scope of philosophical inquiry. Beyond the level of philosophy, the roots of Cartesianism—especially "the method" and Descartes' mind-body dualism—would penetrate all aspects of culture, steadily bringing a new world into being, transforming views about everything from sex to education to the role of women to the relationship between humans and their environment, giving this first modern philosophy fresh life while at the same time pushing its founder farther into the recesses of the past.

3

Unholy Relics

OR A CENTURY AFTER THEIR SECOND BURIAL, IN Paris, Descartes' bones rested undisturbed. But while they moldered and the church of Ste.-Geneviève, in which they were entombed, quietly crumbled into ruin (as it lost its long battles against both the French crown and the neighboring church of St.-Etienne, with which it shared a wall and ancient territorial rivalries), the world of the living transformed itself in unheard-of ways. If someone from any of the previous centuries could have revisited earth in the 1700s, it might reasonably have seemed that human beings had become drunk on invention. Nitrogen was discovered, electricity harnessed, the first appendectomy performed. The income tax came into being. The Hawaiian islands were discovered. The fountain pen was invented, and the fire extinguisher, the piano, the tuning fork, and the flush toilet. Clocks, microscopes, compasses, lamps, and carriages were refined. In the English city of Birmingham alone, the small group of men who called themselves the Lunar Society, epitomizing the passion for combining invention and industry, discovered oxygen, created the steam engine, identified digitalis as a treatment for heart ailments, and built the world's first factories. Men caught in the grip of a mania for collecting and classifying roamed the earth and gathered spiders, minerals, fossils, and flowers. Museums, dictionaries, and encyclopedias came into existence. Surnames—Watt, Fahrenheit, Schweppe, Celsius, Wedgwood—became products or terminology.

To state the above is to state the obvious: the Enlightenment and the Scientific Revolution are part of every school curriculum. What is perhaps less obvious is how threads from these events ran backward in time. We are used to thinking of the Enlightenment as an eighteenth-century phenomenon, in which intellectuals urged changes across society based on a commitment to reason. But recently historians have cleared paths back through the previous century, revealing how networks of events and personalities—inventors and the rippling consequences of their inventions, explorers who wielded sails, microscopes or quill pens—rooted the ideas of Jefferson and Rousseau in the program of Descartes.

The link between Descartes and the ensuing decades of invention and discovery is not so apparent; less still is the link between Descartes and our own era of invention and discovery. Cartesianism isn't exactly in the air these days. There are no university degrees in it. You don't run into parents who say they want their children to grow up to be Cartesians. We tend to prefer "science" when we're talking about a particular systematic way of exploring the natural world. *Science*—the word itself, coming from the Latin for knowledge—has been around since the Middle Ages, when it meant something like art or discipline, as in the science of war or the science of horsemanship. Its use in the sense that we mean today didn't get fixed until around 1800. Meanwhile, "Cartesianism" faded away in the early 1700s.

What happened to the Cartesians is one of the subplots of modernity. In a sense, they were engulfed by a great wave that swept over Europe in the late 1600s and early 1700s, which went by the name of Anglomania. The center of gravity of the "new philosophy" shifted from France to England. This was partly fad, but at its core was an appreciation of English practicality. The French approach to knowledge—ornate, rational, abstract—had elegantly suited the medieval, Aristotelian edifice. Descartes—with his grand system, his effort to construct a holistic view of reality based on reason and the cogito, which would encompass everything from salt crystals to God's grace, from human emotions to Jupiter's moons—was part of that. The British looked at the new thinking as more of a toolbox. They tinkered; they came up with improved metal alloys and ceramic glazes and watch springs. If the French created a new philosophy, the English invented applied science. While the French developed their salons into ornate social institutions, English craftsmen employed apprentices and made them sign hard-nosed contracts such as the one in which young Josiah Wedgwood promised that "at Cards Dice or any other unlawful Games he shall not Play, Taverns or Ale Houses he shall not haunt or frequent, Fornication he shall not commit— Matrimony he shall not Contract."

There was a political dimension as well. Where the French state tried to control the new thinking—in part by creating an academy of sciences that would officially bless or condemn proposed new avenues of study—the English were freelancers. They were thus more adept at innovation, so that inventors in Bristol or Birmingham, acting on their own initiative, raising capital and creating markets, became small-scale industrialists and began to reshape the way the world operated.

The individual most responsible for this change—the person who, it could be said, singlehandedly pushed Descartes into the past—was Isaac Newton. Newton's laws of motion, work in optics, and development of the principles of gravitation formed a hard, practical base on which the scientific revolution would be built. The French themselves lauded Newton as the herald of a new age. Voltaire, the godfather of the French Enlightenment, wrote of *"la supériorité de la Philosophie anglaise"* and praised Newton as "the destroyer of the Cartesian system"—that is, the man who brought science down out of the clouds of theory. This sort of nationalistic divide between thinking and doing crystallized in philosophy departments in the terms *rationalism* and *empiricism*. In this neat compartmentalization, Descartes is not only the father of modernity but the inventor of the "school" of rationalism, which perceives reality from a starting point in the human mind and whose leaders were all continental figures, and that of empiricism, whose main thinkers—John Locke, David Hume, and George Berkeley, English, Scottish, and Irish, respectively—began instead with the reality of the external world.

While there is truth in the shorthand, it is also misleading. Descartes' career, his lifelong focus on medicine and dissection and observation, belies the rationalist label. More to the point, he was foundational to *both* the rationalist and the empiricist traditions, as well as to the Enlightenment's political ideas. Beneath Newton's principles and Voltaire's maxims is the cogito. As a discipline, philosophy itself tends to forget this. As the present-day British philosopher Jonathan Rée puts it, Descartes "was the founder of the 'new philosophy,' whose work was carried on by Newton and later scientists. . . . The principles of the 'new philosophy,' and the theory of knowledge and the theory of human nature which go with it; the concepts of an idea, of mathematical laws of nature . . . are so fundamental to modern consciousness that it is hard not to regard them as part of the natural property

of the human mind. But, in fact, they are a product of the seventeenth century, and above all the work of Descartes."

Thus the essence of Cartesianism—its philosophical kernel,
which encompassed much more than science—not only lived on
but expanded into virtually every corner of human life, evolving
and adapting and spawning new generations, each with its own
characteristic traits but all of them linking back to their ancestor,
even as the original Cartesians flickered into extinction.

"REASON VERSUS FAITH" may be the chronic fever of
modernity, but if the Western world caught it in the period of the Enlightenment the division was not as clear as some
today might like to believe. There seems nowadays to be an in-
grained notion that people of that era set reason firmly against
faith and the two have ever since been locked in a death struggle.
Maybe this idea comes from our desire to simplify things, our
hunger for sound bites and text crawls. Maybe it gives clarity
to both hard-core believers and the antireligion faction, both of
which are very much alive today. People who want to drive society
and politics via the motor of their religious views—whether they
are Muslims, American evangelicals, Roman Catholics, members
of India's nationalist Hindu party—have been particularly vocal
in recent years. But the other side—political atheists, you might
call them—are voicing themselves, too, as evidenced by the titles
of recent books: *The God Delusion, The End of Faith, God Is Not
Great.* The root of these atheist manifestos is the belief that society woke up three or four centuries ago to the realization that
God doesn't control the universe, that rather the blind forces of
nature do, but that many people around the world are still caught

in the trap of religion and are threatening, with violence and intimidation, to drag humanity down the drain. If the hard-core faithful have their ancient texts to rely on for foundations, the new atheists have the Enlightenment.

But the situation was never as simple as that. The fighting was more of a three-way affair, for the new philosophers were themselves split into two camps, each of which would have an enormous impact on modernity and each of which still exists, with representatives continuing their clashes on cable talk shows and in op-ed columns. The split began, as we have seen, with the first-generation Cartesians, with Malebranche et al. adapting the new philosophy to Catholic teaching while Rohault, Arnauld, and others kept the two apart. Over the next generations, the "moderates" continued to believe that reason would function alongside faith to increase human happiness and life span, end disease, reduce suffering of all kinds, and give people greater power over nature and greater freedom in their lives. These moderates worked with the church and within governments: many literally worked in either the church or the state apparatus. The moderate camp includes some of the most well-known figures of the Enlightenment: Montesquieu, Newton, Locke, Jefferson, Hobbes, Voltaire.

Then there was the other element. Two present-day historians of the period—Margaret C. Jacob of UCLA and Jonathan Israel of the Institute for Advanced Study at Princeton—have written nearly identically titled books giving a name to this more shadowy secularist camp. Jacob's *The Radical Enlightenment*, which appeared in 1981, and Israel's *Radical Enlightenment*, published in 2001, focus attention on thinkers of the period who looked to reason as a kind of new faith, who insisted that a necessary object of the thinking that followed from Descartes was to bring about the end of traditional religion—to end what they believed

was the tyranny of superstition in which humanity had existed for
millennia, a tyranny that, they argued, those in power, in church
and state, had maintained for their own benefit. What's more, in
many respects these early Enlightenment radicals didn't just pre-
figure what was to come but fully developed the ideas that would
lead to the world-historic changes of the later era. As Israel puts
it, "It may be that the story of the High Enlightenment after 1750
is more familiar to readers and historians, but that does not alter
the reality that the later movement was basically just one of con-
solidating, popularizing, and annotating revolutionary concepts
introduced earlier."

The changes taking place in the late 1600s and early 1700s
weren't confined to gears and pulleys. Something more than mere
inventiveness was involved. The idea of making reason the ground
to thought and behavior had almost immediate consequences in
the social sphere. As early as the 1660s, the Dutchman Franciscus
van den Enden was advocating a radical new approach to society
that included equal education for people of all classes, joint ownership of property, and democratically elected government. Van
den Enden actually drew up a charter for a utopian community
that would be based in the Dutch New World colony of New
Netherland, with its capital of New Amsterdam on Manhattan
Island. A group of settlers went so far as to establish a base for
the community on Delaware Bay, but within months the English took over the Dutch colony (changing New Amsterdam to
New York) and the scheme—perhaps the first attempt to enact a
society based on the modern principle of democracy—ended. In
the 1720s, Alberto Radicati—an Italian nobleman turned radical
philosopher—similarly argued that natural philosophy showed
that democracy was the only proper form of government. He also
dismissed most biblical teaching and said that people should enjoy

the pleasures of life but that, if life was really awful, suicide was reasonable. (As a true radical, he could only have been pleased when Italy's chief law officer called his principal work "the most impious and immoral book I have ever read.")

Even among the first generation of Cartesians the idea had arisen that, strictly based on reason, there was no justification for social rules subordinating women. Particularly in Paris, women began taking an active part in salons and advancing philosophical discussions. As if to confirm the worst fears of those who criticized such attempts to level society based on sex, erotic literature—novels as well as instruction manuals—began to appear. The thinking behind them, according to one of the most enthusiastic promoters of the new sexual freedom, was that society had denied women the right to understand and express their sexual pleasure as a way to keep them under control. The writer, Dutchman Adriaan Beverland, altruistically devoted himself—both in his personal life and in his work—to freeing women of their sexual inhibitions.

This sexual enlightenment mirrored the path of the wider Enlightenment, with a small number of books in the late 1600s gradually growing into a full literature, of which the marquis de Sade's works—in which sex is a vehicle for exploring notions of radical individual freedom—are the ultimate expression. Indeed, just as much as Jefferson or Rousseau, de Sade was a figure of the High Enlightenment; you might think of him as the Thomas Jefferson of sex. The connection to the new philosophy was also right on the page. Much of the literature of sexual freedom that came out of the post-Cartesian decades had a frankly philosophical cast, with authors undergirding their scenes of women masturbating and cloistered nuns in coital embrace with references to Descartes, Spinoza, Ovid, and Petronius.

As transformative as the threatened sexual revolution promised to be, it was minor compared to the impact of the new philosophy on religious institutions and the religious beliefs of individuals. Before Descartes, religion was the language in which the most basic ideas about life and the world were discussed. Philosophical debates were religious debates: they took place between Catholics and Lutherans, or Lutherans and Calvinists, or Catholicism and Protestantism, or they were doctrinal disputes among members of a particular sect. Beginning around the time of the reburial of Descartes' remains in Paris, the emphasis shifted. Reason applied outside the boundaries of theology—"free thinking"—caught fire and swept across the Continent with a speed and force that bewildered churchmen. As far as the radical philosophers were concerned, Christianity sat on one side of the scale and secular thinking on the other. An English philosopher named Anthony Collins sounded the trumpet of the new thinkers, stating in his best-selling (but anonymous) 1713 treatise, *A Discourse of Free-Thinking:* "By Free-Thinking then I mean, *The Use of the Understanding, in endeavouring to find out the Meaning of any Proposition whatsoever, in considering the nature of the Evidence for or against it, and in judging of it according to the Seeming Force or Weakness of the Evidence.*" And he declared expansively that "if I vindicate Man's *Right* to *think freely* in the full extent of my *Definition,* I not only *apologize* for my self, who profess to *think freely* every day *de quolibet ente,* but for all the *Free-Thinkers* who ever were, or ever shall be."

Philosophers held real sway in the seventeenth and eighteenth centuries. They wrote in newspapers, manned presses and printed their own tracts, thundered in parliaments and councils, debated church leaders, and otherwise molded popular opinion. As a result, the new secularism began to make inroads among ordinary people, and quite soon after the time of the first Cartesians. In the early 1700s, travelers to the Low Countries noted that, as a people, the Dutch seemed to have lost the popular belief in witches and demons; Anthony Collins reported that "the Devil is intirely banish'd [in] the United Provinces, where *Free-Thinking* is in the greatest perfection." The modern French scholar Michel Vovelle studied eighteenth-century archives in southern France and found that starting right around the time of the first Cartesians French people began giving less money to religious organizations and the use of pious language began to drop off in wills and other official writings. Where wills were once replete with pleas to the Virgin Mary and local saints to look after the soul of the departed, by 1750 as many as 80 percent contained no religious references. Of course, Europe remained Christian, but secularism was now a force in society. Gysbert Voetius—who had so vigorously opposed Descartes in Utrecht, saying that his philosophy would lead to atheism and wanton individualism—was right.

In the early 1700s, writers in every European country made names for themselves by advancing the argument that magical thinking—believing in the powers of amulets, in warding off evil, in Satan himself—was nonsense. Some veered toward the forbidden territory of atheism, though almost no one actually espoused it, since professing that God didn't exist was a crime throughout Europe. What arose instead was either deism—belief in God based on reason rather than religious tenets—or "materialistic pantheism," which holds that God and the world, meaning all the physical forces in the universe, are one. Radicati outlined such a view in 1732: "By the *Universe*, I comprehend the infinite Space which contains the immense *Matter*. . . . This *Matter*, modified by *Motion* into an infinite Number of various forms, is what I call NATURE. Of this the Qualities and Attributes are, *Power*,

Wisdom, and *Perfection,* all of which she possesses in the highest Degree."

Power, wisdom, and perfection are, of course, attributes that were formerly assigned to God, and playing fast and loose with definitions in this way did not fool churchmen who were on the lookout for attempts to circumvent their worldview and their authority. In 1708 a German theologian created a guide to enable his colleagues to thwart the kind of thinking that "calls God Nature," which he characterized as "the most systematically philosophical form of atheism."

Most radical Enlightenment figures—Collins, Radicati, Van den Enden, and others—don't have the same star power as moderate Enlightenment players. But not all have sunk into obscurity. Jonathan Israel makes a case that the main force behind the radical wing, its intellectual godfather (and one of the most influential philosophers in history), was Baruch Spinoza. Spinoza used many of Descartes' categories and applied them more ruthlessly—to religion, among other things. Like Descartes, he "proved" that God exists, but he also "proved" that God cannot have human properties and does not perform miracles or otherwise intervene in human affairs. The Bible contains much wisdom, Spinoza wrote, but shouldn't be trusted when it comes to tales of seas parting or water being turned into wine. He ridiculed popular belief in supernatural beings, reacting to a debate about whether spirits can be female or male by saying, "Those who have seen naked spirits should not have cast their eyes on the genital parts."

The three-way debate among the radical and moderate secularists and the theologians ranged over nearly every conceivable issue, but it was centered on the notion of God. The charge of atheism was seemingly constantly in the air in the late 1600s and early 1700s, and not because its targets professed not to believe in God but because they defined God in ways that did not require a church as mediator. A conception of God that did not rest on scripture was considered a danger to church and state. Far from seeing himself as an atheist, Spinoza believed that God *must* exist, for he defined God as infinite substance and reasoned that "a substance consisting of infinite attributes . . . necessarily exists." In his view God was synonymous with nature, meaning not merely the natural world but the totality of all things. He went so far as to upend the medieval notion of substances by defining God as the one and only substance existing in the universe: everything else was some subpart of God.

Spinoza insisted that there is such a thing as religious truth, but he also insisted that religious institutions were largely concerned with protecting their own position. At times Spinoza's thinking about superstition and the manipulation of it sounds not only modern but ultramodern; streamline the language of his *Theologico-Political Treatise* and it could appear in a twenty-first-century antireligion best seller: "The human mind is readily swayed this way or that in times of doubt. . . . Anything which excites their astonishment [people] believe to be a portent signifying the anger of the gods or of the Supreme Being, and, mistaking superstition for religion, account it impious not to avert the evil with prayer and sacrifice. Signs and wonders of this sort they conjure up perpetually, till one might think Nature as made as themselves, they interpret her so fantastically." Religious institutions, Spinoza held, in a passage that has set many people in the centuries since nodding in agreement, prey on this collection of insecurities: "Immense pains have therefore been taken to . . . [invest] religion, whether true or false, with such pomp and ceremony, that it may rise superior to every shock, and be always observed with studious reverence by the whole people."

Spinoza's form of pantheism was reviled by Christians and Jews of his time and later (he was expelled from his Amsterdam Jewish community at the age of twenty-three), but it has fit well in the modern era. Einstein, when challenged to state his own religious beliefs, famously aligned himself with Spinoza, saying, "I believe in Spinoza's God who reveals himself in the orderly harmony of what exists, not in a God who concerns himself with fates and actions of human beings.".

Genuine atheism—a belief that there is no deity involved in the universe or its creation, that we are alone—would, of course, be a major outcome of the modern turn that occurred in Europe in the seventeenth and eighteenth centuries. But it would be wrong to imagine that the Enlightenment was antireligion. Its mainstream thinkers, as well as many if not most of the radicals, were anti-church, not antifaith. Their problem with religion was that it kept individual humans from exercising their own minds and applying their innate reason to understanding the world and their place in it. This criticism applied not only to Catholicism but also to Protestant theology. It's true that Protestantism was a movement on behalf of the individual. It came into being in large part because its leaders felt that individual Christians needed to have their own relationship with God, unmediated by the church. Luther reviled the Catholic Church for making people slaves to the clergy. But at the same time he wrote *On the Enslaved Will,* which argued that individuals must prostrate their intellect and will to the God of scripture. As the marquis de Condorcet, a leader of the French Enlightenment, said of such Reformation figures as Luther, "The spirit that animated the reformers did not lead to true freedom

of thought. Each religion allowed, in the country where it domi-
nated, certain opinions only." The Protestant churches were no
more willing to accept the God-equals-nature argument than was
the Vatican.

The Enlightenment figures wanted people to be utterly free to
use their minds, to apply the light of reason. This included applying reason to faith: evaluating and valuing the underlying substance
of life—the universe, God, nature—with clear eyes, and without
necessarily employing the tools of organized faith. You might say,
in fact, that the whole thrust of the Enlightenment was not an
attempt to diminish God at all but, on the contrary, an insistence
on expanding God, broadening the scope of the word to include
all that the new forms of learning encompassed. The enemies, in
this view, were two: authority—any power or organization that
dictated how and what to believe—and fuzzy thinking.

All of this went directly back to Descartes, whose turn to philosophy began when, after his studies, he found himself "saddled
with so many doubts and errors that I seemed to have gained
nothing in trying to educate myself unless it was to discover more
and more fully how ignorant I was." "Clear and distinct ideas"
would be the goal—of Descartes, and of the thinkers of the next
century. Thus Spinoza lashing out at superstition. In the 1740s,
Denis Diderot, the force behind the famous *Encyclopédie* and one
of the intellectual fathers of the French Revolution, put the zeal
for clarity in the form of a maxim: "Superstition is more harmful
than atheism."

ISTORICAL PERIODS DON'T USUALLY name themselves.
People walking around circa 1300 did not greet each

other with "It's a lovely morning here in the Late Middle Ages."
The Enlightenment—whose leaders were nothing if not self-
conscious—was an exception. *Aufklärung, les lumières, ilustración,
illuminismo, verlichting*: across Europe, in whatever language,
there was an awareness on the part of individuals of somehow
having different minds from earlier generations, and everywhere
they expressed the idea with the metaphor of light invading what
had been darkness. One of the clearest expressions of it came from
the tiny, introverted German philosopher Immanuel Kant, part of
whose grand project was to identify the "transcendental" foundation of religion—to ground faith not in a church or a holy book
but in the human mind, the world, and the relationship between
the two. Kant was a mousy, homebody sort who never strayed
farther than one hundred miles from his Prussian hometown, and
his writings are as dense as any philosopher's, but he could on
occasion rise to the soaring plane of the propagandist. "Enlightenment," he declared when asked to define the force that he and
his contemporaries were caught up in, "is man's exodus from his
self-incurred tutelage. Tutelage is the inability to use one's understanding without the guidance of another person. . . . 'Dare to
know' (*sapere aude*)! Have the courage to use your own understanding; this is the motto of the Enlightenment."

The "motto" was put into practice politically in two dramatic
and very different ways in the eighteenth century. In the 1770s,
there came a metaphorical lull in the frenzy of invention and scientific activity as everyone paused to witness the birth of a whole
new arena of modernity. Across the ocean, the inhabitants of the
former British colonies in North America decided to throw off
their mother country and give a real-world test to the ideas about
representative government that had been nurtured over the century in the theories of men like John Locke. Americans are used

to thinking of their country's revolution as the climax toward which the century of European intellectual ferment was building. As the great American scholar Henry Steele Commager wrote in 1977, "The Old World imagined, invented, and formulated the Enlightenment, the New World—certainly the Anglo-American part of it—realized and fulfilled it." Europeans see it differently. To them, the American Revolution was a sideshow, while the French Revolution, in all its gore and glory and tragedy, its titanic upending of church and state, was the ultimate expression of the Enlightenment and of the long process of transformation that began with Descartes' cogito.

Of course, both revolutions are intimately conjoined to the century of transformation that preceded them. The American leaders—Jefferson, Madison, Adams, and the rest—swam in the same current of ideas the Europeans did. They were steeped in the political philosophy of Locke, who argued that societies are held together by a "social contract" between rulers and the ruled and that if power is abused the people have a right to revolt. Jefferson wrote that his political philosophy derived from Newton, Bacon, and Locke, especially the idea of bringing scientific questioning and observation into the political realm—the cogito, you might say, taking a seat in government. But there was another side to the American revolt. Henry May's 1976 study, *The Enlightenment in America*, argues that religion was so much a part of the American fabric that the issue in America was not "the Enlightenment *and* religion" but "the Enlightenment *as* religion." In the 1730s and 1740s, an evangelical fervor gripped the American colonies; the first so-called Great Awakening gave a spiritual cast to the political drama that would follow. At the same time, many in the American elite were deists, who essentially turned Newton's science into religion. Jefferson made deism part of the nation's fabric with the Declaration of Independence's appeal to "the laws of nature and nature's God." No churches were burned at Bunker Hill or Yorktown. The American Revolution, seen in this light, was the full expression of the moderate wing of the Enlightenment, which stressed order, harmony, and a balance of faith and reason.

But the other Enlightenment, the radical version, would have its own, very different political expression. In 1789, when the people of Paris took to the streets intent on winning a constitution from their king, the scientists and inventors, the cogitators and pamphleteers of Europe paused again and took note, for here was an effort to extend the political ideals of their Enlightenment to one of the old, established nations. Completely unlike what happened in America, the French Revolution was a systematic breakdown of the old order and its representatives, not only the monarchy but the Catholic Church. What had begun in the minds of a small number of intellectuals—antipathy toward both the king and the church for shackling people in cages, for controlling their lives, their minds, their purses—spread to every level of society, with the ferocity and face-to-face, breath-to-breath stench of an unprecedented collision of forces.

If modernity ultimately required a complete break from existing structures—of human thought, belief, society, everything—then what happened in France in the 1780s and 1790s had terrible necessity. It would also demonstrate something that had not been understood by Descartes when he reoriented his worldview around reason but that would become depressingly familiar in the coming centuries. As an organizing principle or a battle cry, reason doesn't necessarily lead to peace and order but can just as well spawn inhuman violence on an epic scale.

It was in the midst of this world-historic lesson that Descartes' bones once again returned to the realm of the living.

❖   ❖   ❖

H E HAD A FACE THAT COULD BEST be described as an-
gelic. If that word suggests purity, it also hints at things
like otherworldliness and ghostliness, and all of these qualities applied to the strange, impassioned, vigorous, meticulous, and ethe-
real man named Alexandre Lenoir. He was born in Paris in 1761,
nearly a century after the remains of René Descartes were buried
in the French capital. In a portrait painted when he was thirty-five,
Lenoir looks creepily like a teenager. His skin shines alabaster, the
lips are mauve and womanish and curled into a slight, freakish
smile, the eyes into which the viewer's are drawn are round and
black like portals onto the dark. On his head is a foppish broad-
rimmed black hat; a gold scarf is knotted around his neck.

He was a lover of art. He studied painting. He married a
painter. He was also, seemingly from birth, obsessively fascinated
by death, images of death, effigies, and human remains. He came
of age in an ideal time in which to exercise such a preoccupation; the French Revolution gave his curious life its context. In
1763, when Lenoir was an infant, King Louis XV levied a series
of new taxes. In the past such a move might have caused only
raucous grumbling, but over the previous decade half of what
would be thirty-two volumes of Diderot's *Encyclopédie* had appeared. Diderot and his coauthors had tried to collect in it all the
new knowledge that was proliferating in Europe, and it proffered
not an objective stance but frank beliefs about, for example, the
connection between a commitment to reason and the moral necessity of obtaining the consent of the governed. Over the years
the volumes, and their underlying logic, had worked their way
into the mental fabric of the country. The parlements of France

banded together to voice opposition to the taxes. Some of the parlements (which were not legislative bodies but rather regional judicial councils) arrested the king's governors; graffiti appeared on
government buildings demanding the king's head. As the matter
escalated, the parlements declared for the first time that together
they represented the will of the people and that taxes could not be
assessed without their consent. The king reacted with a suddenness dramatically out of keeping with decorum. He rode directly
from his palace in Versailles to Paris (pausing at the Pont Neuf
when he came upon a religious procession, before which he dismounted and knelt in the mud as it passed), strode into the Palais
de Justice, and unleashed what has gone down in French history
as the *séance de la flagellation*—the whipping session, one might
say. It was a violent rebuff of the idea of elements of government
uniting in opposition to the head of state, and about as decisive
an assertion of kingly power as is possible to imagine: "In my
person only does the sovereign power rest. . . . From me alone do
my courts derive their existence and authority. . . . To me alone
belongs legislative power. . . . By my authority alone do the officers of my courts proceed."

Thus began the struggle that would lead to the fall of Europe's
most autocratic monarchy. In 1770, the king dissolved the parlements, but in a sense the damage had been done. Pamphleteers
had broadcast the reasoning of the parlements; it resonated with
the people and continued to do so through the following years.
In 1788, a restored Parlement of Paris warned a new King Louis
(the XVI) that they would not stand for royal despotism; now
the parlements were using phrases like *rights of man* and *confirmed
by reason*. The following year, representatives of the Third Estate (going back to the Middle Ages society was divided into a
First Estate, comprised of the clergy, a Second, the nobility, and

a Third, commoners) took the new language further, declaring that they were not an estate at all—not a third-rate advisory council to those in power—but "the people." In fact, they were a National Assembly. The king locked them out of the meeting hall. They gathered instead at a nearby tennis court and took an oath to remain united until the king agreed to a constitution. Soldiers marched into Paris. In one stroke, the National Assembly "abolished feudalism," then issued a Declaration of the Rights of Man and Citizen.

Alexandre Lenoir grew to maturity during the tumult leading up to the Revolution, and the waifish artist was part of the fervor. And yet his interests soon parted in one crucial way from the objectives of the revolutionaries. His mentors included the artist Gabriel François Doyen, who made a name for himself with florid large-scale religious paintings in the manner of the Italian Renaissance, and the polymath Charles-François Dupuis, who besides being one of the inventors of the telegraph wrote a wildly successful book called *The Origin of All Religious Worship*, in which he argued that Christianity was merely an updating of ancient cults of sun worship. Bathing in the lush influences of Doyen and Dupuis, as well as Freemasonry (which itself sprang from the minds of freethinkers as a ritualized theologizing of nature), and steeped in his own exotic personal mysteries, Lenoir developed a private universalist belief system that centered on reason, history, religious art, and architecture.

To his horror, as the Revolution escalated mobs of his fellow revolutionaries took literally the calls from their leaders to tear down the structures of the old regime. Crowds attacked churches and palaces. Buildings were looted, paintings and sculptures destroyed; monks' cloisters became stalls for horses. One by one, many of the country's most ancient religious structures—the Abbey of Cluny, the church of St.-Denis, burial site of the French monarchy—were ransacked. Once-precious relics, including the bones of formerly revered kings, were paraded through the streets. The body of Louis XIV himself, still in a state of semi-preservation, was unearthed and hacked with knives, to cheers. It was madness, but with a method. The developing ideology of the Revolution emphasized the values of liberty, equality, and fraternity, with their roots firmly embedded in seventeenth-century "new philosophy." This ideology rejected all symbols of the past that put "mysticism" over reason, that "did not bear the stamp of utility," and that were "contrary to good morals." The government not only egged on crowds but carried out an official "dechristian-ization" program that resulted in the destruction or desacration of religious buildings in virtually every town in France, from village churches to Notre-Dame de Paris, where the sculptures of biblical figures on the façade were defaced.

The revolutionary leaders were not entirely rabid in their zeal, however. As the ideological vandalism escalated, some people within the revolutionary committees fretted. In fact, it was in a report from the Committee of Public Instruction to the National Convention despairing the loss to the country that the word *van-dalisme* was coined, referring to the fifth-century Germanic tribe that became infamous for sacking Rome.

It happened that the painter Doyen was on one such commit-tee. Lenoir, his disconsolate pupil, approached him with an idea. What if the government, while not rescinding the call to tear down the old order, nevertheless chose someone to sort through the revolutionary debris for works of art that might have historical value? Surely there was a balance to be found between destruction of symbols that carried the poison of slavish obedience to king and church and the obliteration of a nation's memory.

Doyen brought the idea to the mayor of Paris, who in turn presented it to the revolutionary government. Perhaps to his own astonishment, Lenoir found himself being offered the job of making order out of the artistic and architectural disorder of the revolution. He was given a broad mandate, two assistants, and a salary. A location was chosen as a repository for the items he saved for preservation: one of the Catholic holdings that had been commandeered by the government, the former convent of the Petits-Augustins, on the bank of the Seine.

Lenoir set to his work with (take your pick) religious or revolutionary zeal. Word would come to him of an assault on a monastery or church or chateau; through the scarred streets of war-torn Paris he and his assistants would rush. Arriving at the scene, he would brandish a writ from the Committee of Public Instruction or the Committee of Alienation of National Goods, demanding, in the name of the revolutionary government, that certain items not be harmed. The crowd would fall back; Lenoir and his men would haul the spoils into wagons and transport them to his depot on the river. Part of a weekly log ran as follows:

> Wednesday.—An angel from the tomb of Bérulle; the mausoleum of Louvois.
> Thursday.—Marbles from the Oratory of the Capuchins.
> Friday.—A *Cybèle* and a *Méléagre*.
> Saturday.—A philosopher, in the antique manner.
> Sunday.—The statue of Cardinal Bérulle from the Oratory of St.-Honoré. The statue of Cardinal de Richelieu from the Sorbonne.

At times revolutionaries refused to acknowledge Lenoir's official sanction. During the last-named appropriation, he struggled with soldiers who were in the process of destroying the tomb of Cardinal Richelieu and was wounded in the process. An engraving shows him similarly fending off sansculottes armed with pikes and axes as he protected the tomb of Louis XII.

Of the hundreds of religious sites seized by the state during the Revolution, one had a unique story. The church of Ste.-Geneviève, dedicated to the patron saint of Paris and occupying the highest point in the city—which also happened to be the burial place of René Descartes—had been in poor condition as far back as 1744, at which time Louis XV made a vow to build a new church. His architect, Jacques-Germain Soufflot, a lover of classical Greece and also of Gothic design, conceived of the new building along the lines of a Greek temple, with a massive portico of columns in front and a soaring central dome, as well as medieval elements. It took decades to create the structure, which went up just across the plaza from the old church. As the Revolution broke out it was finally nearing completion—just in time to be condemned by the revolutionary government as a temple of feudalism and mysticism, two isms that were nearly equated with evil.

The building was taken over by the state and, perhaps partly because its spare classical structure fit with the artistic ideals of the Revolution, was dechristianized and converted into what is arguably the purest expression of the "radical" Enlightenment in stone: the Pantheon. The Pantheon was dedicated not to gods in the usual sense but to the great men of France—or, as it was phrased in the histrionic spirit of the time, "to fame." Like a church, it would be a place of reflection, but this would be a hall devoted to rigorously secular reflection. Like a church, it would house human remains—but rather than indicate a link between the physical body and the immaterial soul it would sing the connection between physical remains and the great work done on earth by the living, work in the service of the march toward human freedom and equality. It would be a secular temple, a shrine to human reason and human progress, stripped of religion and "superstition."

In redesigning it so, architecturally replacing faith with reason as a source of worship, the revolutionaries created a unique monument, and visiting it today gives a feel not only for their motivation but for its naïveté and hollowness. The strangeness comes sweeping over you the moment you enter: the vastness is almost as laughable as the idea of dedicating a building to "great men" and "fame." It sounds lampoonable, vacuous. Scenes from myth and French history are painted on the walls, but there is nothing in between. It's yards and yards of empty space, with columns standing like trees in a desert of marble. And downstairs, in the crypts, the tombs of the great men (and, these days, a few women) are lit with such dramatic relief you would swear you were on a film set. Maybe the oddest thing is the unyielding lack of adornment, the painstaking absence of religious motif in a sanctuary devoted to the dead. In a place like this the idea is driven home to you that reason alone is an empty vessel.

The secularization of the building, its associations with science and order, extended in a variety of directions. The cross at the top of the dome was replaced with a globe (which was later replaced by a statue and ultimately a cross again). The top of the dome, being the highest point in the city, also served as a platform from which, as the Revolution raged on, Jean-Baptiste-Joseph Delambre, the great French astronomer, did his calculations of the earth's size to determine the new, scientifically based unit of measure—the meter—that would replace the old feudal mishmash of systems. The building's attraction as a hall of modernity would continue into the nineteenth century, when the physicist Léon Foucault would hang his pendulum from the top of the dome, offering proof of the rotation of the earth and giving thousands of people who came to see it a tangible sense of science.

The Pantheon symbolized an entire approach to modernity. Dechristianization became an official component of the modernizing program of the French Revolution, and it spread into every facet of life. The calendar—built around the Catholic holy days—was scrapped in part for being tainted by religion, and a new one was created based on the scientific observation of nature: the sun, the moon, the season, the turning and revolution of the earth. Names of streets and towns were stripped of religious associations: St.-Jacques, St.-Louis, and St.-François vanished, while republican heroes both local and imported—not just Danton and Mirabeau but Cato, Brutus, and Benjamin Franklin—sprang up on signs and maps. So thoroughly did the Revolution equate religion with superstition that as it moved into its most radical phase a ceremony was held at Notre-Dame cathedral in which religion was denounced, atheism was proclaimed, and a "Cult of Reason" was declared, with an actress playing the role of Liberty prostrating herself before the burning fire of Reason.

To appreciate the difference between the French and American revolutions, you might try to imagine the American founding fathers, egged on by a mob, stripping a church in Philadelphia of its religious overtones and converting it into a temple of reason. That such an image seems preposterous makes a point not only about history but perhaps also about present-day realities. At least some of the problems that the Western world confronts today, as it grapples with such forces as militant Islam, have to do with the fact that the modern Western world has a split personality: it is confused and divided over the relationship of reason and faith, whether there can be a relationship or whether the one supplants the other. In simplistic form, the United States, where religion is still a strong force in both public and private life, maintains the moderate Enlightenment tradition—a moderate modernity—and Western Europe, which has largely abandoned organized Christianity, has tended to follow the radical path. That split runs straight back to the difference between what happened in 1789 and what happened in 1776—and, of course, back to Descartes.

While the new Ste.-Geneviève was being radically secularized, the old, ruined church still stood. It contained dozens of tombs, monuments, and statues of saints, and thus, in 1792, it became a target of revolutionary zeal. After the government closed it to religious worship, its abbot appealed to the guardian of the *"dépôt des monuments"*—Lenoir—to save what could be removed before the building was destroyed.

Among the tombs and monuments, of course, was that of Descartes. One hundred and forty-two years after his death, Descartes was now seen as part of France's legendary heritage. He was a "great man," and the revolutionaries—some of them—wanted to recognize him as such. The year before Lenoir was asked to rescue the remains from the old Ste.-Geneviève, Descartes' bones became part of a debate in the revolutionary government over the conversion of the new church into the Pantheon—and, more than that, over what the Revolution itself meant. The building was to be a temple to greatness, but who was great? How to decide whose remains would be given the honor of "pantheonization"? The debates, carried on in the midst of the upheaval and vituperation of the Revolution, were at times heated to the point of viciousness, and the history of the first few "great men" to be selected—and deselected—for pantheonization says a thing or two about the flaws inherent in the idea that reason could be a purely objective force.

It was the death of Honoré Mirabeau, one of the universally revered heroes of the Revolution, on April 2, 1791, that gave rise to the idea of turning the new church into a secular mausoleum, so it seemed at first that the factions within the National Assembly would unite in support of the pantheonization of Mirabeau himself. But shortly after he was buried in state in the bowels of the vast building it was discovered that he had been making concessions to the monarchy as the king sought to stay in power (and stay alive). Robespierre argued that Mirabeau had betrayed revolutionary ideals and should be disinterred and removed from the Pantheon. He was—quietly, embarrassingly, by a side door.

While that debate was being carried on, the leaders united around the idea of moving the remains of Voltaire into the Pantheon. It was a popular choice—one of the very few that came without rancorous debate. Voltaire had at times been reviled as an atheist; he had been imprisoned in the Bastille. Now he was considered the (secular) patron saint of the Enlightenment. On his death thirteen years earlier he had been buried quietly, but the procession to rebury him in the Pantheon became one of the great events of the Revolution. One hundred thousand people lined the streets to see the parade, which included a full orchestra pulled by twelve white horses, a gold box bearing the complete ninety-two-volume library of Voltaire's literary output, a phalanx of ordinary *citoyens* who had proudly taken part in the attack on the Bastille, and a flag-draped, triple-tiered sarcophagus containing the remains of the man himself.

But Jean-Paul Marat—the revolutionary who was one of the forces behind the Reign of Terror and who is best remembered today for the painting of him by Jacques-Louis David, slumped dramatically dead in his bath after being knifed—suffered a fate similar to Mirabeau's. Barely a year after he was chosen as one of the new republic's secular saints, sentiment swung against him, and his remains were hauled back out of the Pantheon.

On April 12, 1791, just ten days after Mirabeau's death launched the matter of pantheonization, the National Assembly had the case of Descartes put before it. The original petition was brought forth by a descendant of Descartes' elder brother, Pierre, but it was taken up by Condorcet, one of the leaders of the Revolution and, along with Voltaire, one of the men whose work most fully embodied the spirit of the Enlightenment. The several hundred men ranged in a circle of tiered benches around him in the central government chamber all knew of his wide-ranging energies. Condorcet had for years been active in working to bring a scientific perspective to bear in politics, economics, and education, to reform all of society around the principle of reason. As a mathematician, he identified what became known as Condorcet's paradox, a mathematical discrepancy in majority rule voting. Politically, he was an unusually early proponent of total equality who argued in favor of granting women and minorities full rights.

Condorcet identified the source of the great change that he and his contemporaries were living through. In the century before, he wrote elsewhere, Europe had been in "the shameful slumber into which superstition had plunged her." It was Descartes who "brought philosophy back to reason," for "he had understood that it must be derived entirely from those primary and evident truths which we can discover by observing the operation of the human mind." Now Condorcet put before his fellow revolutionaries the case of one of their countrymen who had been more fundamental even than Voltaire in leading to the extraordinary events in which they themselves were now participants. "Descartes, who was forced from France by superstition, died on foreign soil," he began theatrically. "His friends, his disciples, wanted at least that he have a tomb on his own soil. His body, transported through their cares, was deposited in the old church of Sainte-Geneviève. . . . They had prepared a public elegy, but superstition forbade praise to a philosopher, pride did not allow honor to fall to an individual who was merely a great man. . . . But this long wait can perhaps be repaired. By breaking the bars restraining the human spirit, he prepared the eternal destruction of political constraint, and deserves to be honored in the name of a free nation."

Condorcet made a persuasive case. The assembly agreed to send the petition to the Committee of the Constitution. But events on the ground threatened to overtake the politicians. While the committee was considering moving Descartes' bones from the old Ste.-Geneviève across the plaza to the new Ste.-Geneviève (that is, the Pantheon) Lenoir got the news from the abbot of the old church that it was in the process of being ransacked; he was appealed to rescue its precious objects. This was a particularly chaotic stretch of months in the life span of the French Revolution. Fears (and expectations) that the Revolution would spread beyond the country's borders led to the outbreak of war with Austria in April 1792. In August radicals took over the Paris Commune and pushed the national body to revoke the king's powers once and for all and usher in a true republic. Dechristianization reached a climax in September, when crowds, convinced that Catholic priests were undermining the revolutionary effort, stormed Paris prisons (priests who had continued to say Mass having been rounded up in the previous months) and murdered 230 priests and more than 1,000 other prisoners. That same month the monarchy was dissolved; in January, the thirty-nine-year-old Louis XVI—bookish, earnest, regal, out of his depth—had his head removed from his body, and the Revolution was well and truly consummated.

Lenoir collected feverishly through all the upheaval and carefully logged the items he gathered:

> From *St.-Etienne-du-Mont*, the epitaph in white marble of Blaise Pascal . . .

> From the church of *Notre-Dame*, two kneeling statues, by Coustou and Coyzevox, representing Louis XIII and Louis XIV . . .

> From *St.-Chaumont*, a statue, in plaster, of *la Ste.-Vierge;* another of *St.-Joseph*, and a bas-relief representing *Jesus in the tomb*, also in plaster, by Duret.

> From *St.-Benoît*, the epitaph in white marble of Winslow, the celebrated anatomist.

He also later admitted that it was personally an unhappy time for him—which suggests that he was not quite himself—though we don't know what the source of his unhappiness may have been. Still, over the course of several trips that he and his assistants made to the old Ste.-Geneviève he meticulously recorded numerous items that he detached from the church and hauled back to the safety of his storehouse on the Seine:

> Four female figures, sculpted in wood, by Germain Pilon . . .

> The reclining statue, sculpted in stone, of King Clovis I . . .

Two columns from Verona; two more in Flemish marble . . .

Two small columns in gray granite . . .

The kneeling statue in white marble of Cardinal de la Rochefoucauld, accompanied by an angel holding the back of his coat . . .

Two ancient black columns coming from the lower chapel. . . . A black marble table that supported two terra cotta works by Germain Pilon, representing "Jesus in the tomb" and "his Resurrection". . . . Two other small columns that also came from the lower chapel.

Lenoir made a careful diagram of the layout of Ste.-Geneviève showing the placement of every coffin that lay beneath its floors, as well as meticulous sketches of many of the coffins, some of which included gruesome renderings of corpses in states of semi-preservation.

In addition, on New Year's Day of 1793, he took pains to record an accident. He had given his assistant, a carpenter named Boucault, the job of removing the "richly ornamented" marble tabernacle of the church. Boucault hauled it onto a sleighlike structure that was harnessed to eight horses, which would pull it to the depot, but when the horses strained the sleigh collapsed and the tabernacle broke into pieces. So meticulous was Lenoir in his tallying that he even indicated how he disposed of the fragments: by selling the copper-gilt bases and capitals to the Hôtel de Nesle in the rue de Beaune.

It's curious, then, that with so much careful attention given to Ste.-Geneviève he failed to record the retrieval of Descartes' remains in his log. He later insisted that he did indeed dig up the philosopher's grave. Not only did he bring a container of Descartes' remains—bones and bone fragments—down the hill from the church to his repository by the river, he declared, but he was particularly excited about its contents. Like Condorcet and other Enlightenment figures, Lenoir believed in the idea of progress: that with each generation, each passing century, humanity was evolving upward, toward happiness, freedom, equality, a higher state of civilization. The current generation, and in particular the Revolution in France, was the point toward which all of Western history had been evolving. For Lenoir, Descartes was not only one of the prime movers of history, he was "the father of philosophy" and "the first to teach us how to think."

Lenoir's fascination with death and bones and graves went deep, so to speak. Along with rescuing tombs and monuments, he unearthed many human remains, including those of some other notable historical figures, including Molière, the famous medieval lovers Heloise and Abelard, and Descartes' protégé Jacques Rohault. There are indications that Lenoir took to this aspect of his work with eerie relish. One colleague recollected him breathing in the perfume of a freshly opened coffin and reverently plunging his hands into its dank innards. After unearthing the corpse of King Henri IV he delighted in its excellent condition, writing, "I had the pleasure to touch these pleasant remains, his beard, his reddish moustache so well preserved." He couldn't stop himself from shaking the monarch's petrified hand despite the fact that, as he felt the need to reaffirm, "I was a real republican." In the case of Descartes, he later wrote that he went so far as to take a fragment of bone—"one very small piece of bone plate"—out of which he carved several rings. That is, Lenoir fashioned a bit of Descartes into *jewelry*. These rings, he wrote, "I offered to friends of the good philosophy."

Louche and creepy as this would have been both to someone from Descartes' era and to someone of our own, Lenoir was not alone; the use of human bones and hair as trophies, ornament, tchotchkes was a feature of his day, a secularization of the Catholic cult of relics. As the Pantheon itself showed, the modernist need to distance society from religion didn't obviate the human need to connect with the past, to come to terms with mortality. Just as religious buildings were co-opted for secular, humanistic purposes that were nevertheless somehow transcendent, the notion of certain human bones becoming conduits between the mortal and the divine was taken over and given new meaning. They may have been desacralized, symbolic of worldly achievement and advance, but the Enlightenment still had its relics. What's more, the fetishizing of remains continued into the next generations. The nineteenth-century explorers who roamed the earth in search of specimens of one type or another assembled them into "cabinets of curiosities" with which to decorate homes and impress visitors, and these often included bits of famous somebodies. A supposed piece of Descartes' skull sits today in the collection of the Historical Museum of Lund, Sweden, where it originally formed part of such a cabinet.

Lenoir later said that when he dug in the church he found the remains of Descartes in a rotted wood coffin, so back at his depot he carefully transferred them to what he considered a fitting and permanent home: an ancient Egyptian sarcophagus made of porphyry that he had taken from the church of St.-Germain-l'Auxerrois. He placed it in the garden of the former convent, alongside his growing collection of statues and tombs—ultimately, thousands of artistic stone objects from all over France, dating from nearly every period of the country's history.

Meanwhile, the government committee considering pantheonization issued its report, which was promptly taken up by the entire body, and it is grounding—humbling, even—in an era that values speed over reflection to think of the revolutionary government pausing at length in the middle of its momentous work to consider such a matter—stopping history, in effect, in order to ponder its course and their place in it. The report was presented by Marie-Joseph Chénier, a playwright who had endured failure after failure until one of his plays with revolutionary overtones was mounted shortly after the storming of the Bastille and he became an overnight sensation—and then a member of the revolutionary government. His brother André was even more famous as a poet of the Revolution. In the course of studying Condorcet's proposition, Chénier became enamored of the idea of Descartes as the first champion of reason, liberty, progress, equality—in short, as the father of the Revolution. Chénier was young, handsome, fearless, impassioned, and at present he and his brother were darlings of the Revolution (the following year André would succumb to the guillotine); he delivered a flowing discourse on behalf of the committee:

Citizens,

Your committee of public instruction has charged me to put before you an object that concerns the national glory and that offers you a new occasion to show to the eyes of Europe your respect for philosophy, the source of valid institutions and true popular laws. In the first centuries of the French empire, a villager from Nanterre was declared holy and was proclaimed patron of Paris. Today, Paris and all France have only Liberty as a patron. A temple was built to Geneviève: this temple, now as outdated as prejudice, is collapsing under the hand of time; but amid this religious rubble, near the sacred relics that, through the follies of man, the pious beliefs of our ancestors imbued with a sterile trust, amid altars enriched by fear, among tombs ornamented by pride, a narrow undecorated stone covers the mortal remains of René Descartes.

It is of course conceivable that in the confusion of the times Chénier and his committee were unaware that Lenoir had already retrieved those mortal remains from the old church. There is another possibility, however, which we will consider later. Chénier went on to place Descartes at the forefront of the line of thinkers whose work had formed the backbone of modernity—"Locke and Condillac, . . . Newton, Leibniz, Euler, Lagrange"—and summarized his committee's finding: "We have thought that a nation that becomes free through the beneficial effect of the Enlightenment must collect with veneration the ashes of one of its prodigious men who advanced the scope of public reason." He decried the wayward life of the philosopher, who was forced by "despotism" to wander Europe, and concluded with a flourish: "To you, republicans, belongs the task of avenging the contempt of the kings for the remains of René Descartes."

The government agreed, and crafted a decree:

DECREE
OF THE
NATIONAL CONVENTION
Dated the 2 and 4 October 1793, the second year of the
one and indivisible French Republic,

*Who accord to René Descartes the honors of the
great Men, and order the transfer of his body to the
French Pantheon, and his Statue made by the
celebrated Pajou.*

1. On October 2.

THE NATIONAL CONVENTION, after hav-
ing heard the report of its Committee of Public Instruction, decrees the following:

ARTICLE ONE.
René Descartes has merited the honors of the great men.

II.
The body of this philosopher will be transferred to the
French Pantheon.

III.
On the tomb of *Descartes* will be engraved these words:
In the name of the French People
The National Convention
To René Descartes
1793, second year of the Republic.

IV.
The Committee of Public Instruction will consult with
the Ministry of the Interior to fix the date of the
translation . . .

2. On October 4.

The National Convention decrees that the statue of
Descartes, made by the celebrated *Pajou*, and which
is found in the hall of antiques, will be removed to be
placed in the Pantheon the day when the remains of this
great man will be transferred there; and authorizes the
Ministry of the Interior to make all the arrangements
necessary to carry out this work.

It was a grand, official, full-on acknowledgment not only of
the place of Descartes in French history but of the forces at play
in history and of the idea of progress. It was in a way a perfect
moment for society to acknowledge these forces—but in another
sense it may have been too perfect, for the Revolution was reaching its bloody summit. The monarchies of Europe—in league with
aristocrats and churchmen inside France—were trying to stop
the dangerous revolt against the political status quo, which had
led to a succession of wars against the revolutionary regime. The
wars and intrigue in turn worsened living conditions among the
people, and hunger inclined the populace toward the most radical
elements among the revolutionaries. Robespierre took control of
the government and introduced institutional terrorism as a way
to deal with perceived threats to the new republic. The guillotine
became a symbol of the bloody excesses of the Revolution. Heads
rolled—tens of thousands of them—including, most famously,
that of Marie Antoinette and ultimately that of Robespierre himself. More pointedly, Condorcet, too, had become ensnared in the
Terror. As a result of certain unrevolutionary tendencies—he had
opposed the execution of the king, for one—a warrant was issued
for his arrest the day after Chénier gave his talk in support of
Condorcet's appeal for pantheonization, and he was forced to flee.

It was while he was in hiding that he wrote his *Sketch for a Historical Picture of the Progress of the Human Mind*, the book in which he summarized his belief in the Enlightenment and its values and in which he singled out Descartes' contribution. He was eventually captured and died in prison, under dubious circumstances.

Thus, if the pantheonization of Descartes during the French Revolution—arguably modernity's most sharply honed act of self-expression—was fitting, the fact that the decision to honor modernity's founding father came on the eve of the Reign of Terror was doubly so. Liberty, equality, democracy—all were offspring of the cogito and the orientation of humanity around reason. But already in 1739 the Scottish philosopher David Hume had argued that it was a mistake to think that reason is the basis of moral principles: reason, he knew, could be put to the most unreasonable pursuits. As a tool it can build a new society, but it can also kill and maim, and misusing it—through naïve belief or duplicity—is one of the tropes of modern history. Historians have long looked at the Reign of Terror—the state's suspending laws and putting violence to official use for supposedly noble and rational purposes—as the forerunner of many recent evils, from Stalin's purges to the infamous "We had to destroy the village to save the village" logic of the Vietnam War. " 'Tis not contrary to reason to prefer the destruction of the whole world to the scratching of my finger" was how Hume derisively characterized reason's negative application. In the very same year that the Reign of Terror broke out in France, Kant, from the isolation of his German village, pondered the conundrum that while reason was now identified as the first principle of modern society, humanity's "propensity to evil" was undeniable. His conclusion is still our conclusion:

Man himself must make or have made himself into whatever, in a moral sense, whether good or evil, he is or is to become. Either condition must be an effect of his free choice; for otherwise he could not be held responsible for it and could therefore be morally neither good nor evil. When it is said, Man is created good, this can mean nothing more than: He is created for good and the original predisposition in man is good; not that, thereby, he is already actually good, but rather that he brings it about that he becomes good or evil, according to whether he adopts or does not adopt into his maxim the incentives which this predisposition carries with it.

Just as it was grimly appropriate that Robespierre and other instigators of the Terror themselves fell victim to it, it fits modernity's taste for irony that the purges and violence in 1793 and 1794 derailed the effort to pay homage to one of the progenitors of the modern world. With so many fresh bodies to bury, with so many of its own members facing arrest or execution, the National Convention found that it had more pressing things to do than fix a date for the transfer of one decayed set of human remains. Descartes' bones stayed in Alexandre Lenoir's depot.

THE TERROR, HOWEVER, WAS good for Lenoir's business. As the mayhem mounted, so did his collection. It elegantly littered the interior and the grounds of the lovely former convent by the Seine, creating a sonorous cascade of sculpted and chiseled effects, beauty alternating with violence, the stately flow of the past jumbled by a deforming present. It was at this time of world-historic change and chaos that Lenoir got an idea. What, eventually, was to be done with the glorious rubble—with all of these mementos of a nation's past—that he had salvaged? Again, *progress* was uppermost in his mind. He thought of past eras as building to the current moment—the age of reason and enlightenment. Those eras were represented and reflected in their art. Was it all to fade? Faced with the destructive forces that were upon them, would people forget it all? Memory of the past could be wiped away in a generation—would that not be terribly wrong? Should history, its lesson, its forward march, not be imprinted on the minds of citizens of a democracy?

What if he were to bring the force of reason—the same force that had brought about this violent wholesale change—to bear on this dislodged material, impose order onto it, give people a clear representation of history as having an underlying purpose? His idea was to create a space for history and art, an educational forum, a place that would show humanity's noblest sentiments at work. He would create a temple to the muses: a *museum*.

A striking thing about the people who handled Descartes' bones through the centuries is how nearly all of them embodied in some way one of the aspects of modernity that Descartes is credited with bringing into being. When, in 1796, with government backing, Lenoir transformed his depot into the Museum of French Monuments, he created possibly the first-ever history museum and became one of the first people to bring a social-scientific approach to art and history. Using reason and progress as guiding principles, he took debris from the destruction caused by revolutionary war and built something new: a public institution that told the story of the nation and its evolution.

There was already a national museum under way, and the minister of the interior stipulated in his letter of authorization to

Lenoir that the institution would technically be a branch of the new Louvre. Lenoir bristled—the Louvre, in his estimation, was a mishmash. He was determined that his museum would have an organizing principle.

And so it did. His first determination was that the visitor would experience history as a progression from lower to higher orders of civilization. In walking through the museum, one would move from century to century, chronologically following the advance. As he worked, he exhibited a great flair for design, and he gave each room an atmosphere he thought suited its historical era—as well as its own funereal aura. He described the first room, devoted to art of the thirteenth century, this way:

> Sepulchral lamps hang from the vaults. The doors and windows . . . were designed by the celebrated Montereau according to the taste of the architecture revived by the Arabs. The window glass also bears the stamp of that style. . . . The somber light that pervades this hall is also an imitation of the time . . . [representing] the magic by which men maintained in a perpetual state of weakness human beings whom superstition had struck with fear.

Lenoir's views about history and progress show up in his description of the use of light in churches of various eras: "I have observed that the farther one goes toward the centuries which approach our own, the more the light increases in public monuments, as if the sight of sunlight could only suit educated man." Until, presumably, one gets to the Revolution, when roofs were literally ripped off churches, exposing the dark interiors to the full light of day.

Lenoir's Museum of French Monuments marks not only the beginning of museums but the beginning of a familiar complaint with museums: that they dislocate objects from their source and purpose and original meaning and force them into a new, alien structure. Museums squeeze new meanings out of objects, ones their creators never imagined. The carved Virgin that for centuries stood next to an altar in a Provençal village church, to which generations had prayed—so that it was *their* Madonna, an object that blended their reverence for a woman of first-century Palestine with all the heartfelt and commonplace aspects of their daily existence, an object that was as much a part of their lives as the mountains framing the landscape—now occupied a wall alongside other dissociated items from roughly the same century and helped to tell the story of the development of realism in art, which, for Lenoir, showed the evolution of humanity.

If this is a complaint of modernity, of modern life, of the force of reason—that it takes things out of the organic pattern in which they evolved, breaks them into analyzable bits, reconstructs them in new ways that may shed new light but that, for many people, have a chilly, inhuman glow—it's all the more interesting that Lenoir's museum became the single most popular cultural site in Paris in the years of the Revolution. Strange to say, tourists actually came to the city in the midst of the upheaval, and the Museum of French Monuments was on everyone's itinerary (an English *Sketch of Paris* published in 1801 devoted fourteen pages to it), so much so that Lenoir published a catalog, which was later translated into English, which people bought (for five francs) and strolled with as they conducted themselves through the sepulchral gardens and rooms. Lenoir began the catalog by trumpeting the underlying theme of the project: "The French cherish this famous revolution that took place through them and by them. This revolution established a new order of things founded on reason and justice." The catalog also contained, in front, a notice of such impeccably humdrum practicality that it could serve equally to signal modernity: "The Museum of French Monuments is open to the public Thursdays from six o'clock to two and Sundays from six until four in the summer and until three in winter."

As concerned as he was to instruct the public, Lenoir also organized his museum around his macabre tastes. At its center was a garden filled with historic tombs. This was his pride and joy, a *jardin élysée*—named for the arena of the afterlife in ancient Greek mythology reserved for the noblest souls—in which a visitor was meant to ponder beauty and death. His description of it in his catalog shows his special savor for things sepulchral: "In that calm and peaceful garden one sees more than forty statues; tombs set here and there on a green lawn rise with dignity in the midst of silence and tranquility. Pines, cypresses, and poplars accompany them; death masks and cincrary urns placed on the walls combine to give this pleasant place the sweet melancholy which speaks to the sensitive soul."

The importance of the *jardin,* according to Lenoir's sensibility, lay in its concentration of the bones of distinguished men and women of the past—philosophers, poets, painters, playwrights— who contributed to the glory of France. His belief was "that their reunion in one place only concentrates that glory in order to spread it abroad with even greater brilliance." Contemplating this public "reunion" sends him over the top in his reverie:

> May one imagine these inanimate remains receiving a new life, being seen and heard, enjoying a common and unalterable bliss? Is the picture of the antique Elysium more seductive than that offered us by such an imposing gathering? . . . I am pleased to say that I feel a new and sweet emotion every time I step into this august enclosure; I would add that the reward

dearest to my heart would be to pass on to the souls of my readers and those who visit this *élysée* the holy respect with which
I was imbued, while creating it, for the intelligence [of those
resting here], for their talents, and for their virtue.

The garden was where Lenoir placed the stone coffin containing Descartes' bones, which he described (and duly numbered) in
his catalog:

No. 507. Sarcophagus, in hard stone, and hollowed in its interior, containing the remains of René Descartes, died in Sweden
in 1650, supported on griffins, an astronomical animal composed of an eagle and a lion, both sacred to Jupiter—and the
emblem of the sun, which represents the home. The poplars,
which climb nearly to the top of the clouds, the yews, and the
flowers shade this monument, erected to the father of philosophy, to him who was the first to teach us how to think.

But how long would Descartes' bones rest in the shade of the
yews and poplars? France now had yet another new revolutionary government—the Directory, in which five directors formed
the executive branch, which governed with two legislative chambers—and just as Lenoir's museum opened its doors, the Council
of the Five Hundred, the lower house of the newly formed legislature, took up the matter of pantheonization once more. Again,
it's remarkable, sitting today in the postmodern cloister of the
National Library in Paris, poring over the original pages of the
council's legislative record—weathered, sepiaed, spotted with
mold, the alternating of bold Roman serifs and plaintive ital-
ics typographically signaling the charged times—to realize that
in the midst of so much vital activity the legislators could be-

come completely absorbed in a debate about so seemingly arcane a topic. Over a period of a few days the council debated the status of refugees, the matter of "defendants charged with assassinations and massacres committed at Lyon and in the departments of the Rhone and the Loire," property taxes, "the conservation of our manufacture of silk, linen, and wool," "the reestablishment of officers of the peace in Paris," and "the means to vivify the public spirit."

In the midst of which, on May 7, 1796, Marie-Joseph Chénier once again addressed his colleagues. The matter was supposed to be simple—finally carrying out the order to transfer the remains to the Pantheon—but so symbolic an event had now become politically charged. *"Citoyens répresentans,"* he began, using the revolutionarily correct form of address, "the remarkable question that your commissioners were called to examine and that the legislative body has to resolve today, relative to René Descartes, is to know if the translation of his remains to the Pantheon should take place the 10th of Prairial, the day of the *Fête de la Reconnaissance,* conforming with the invitation made to you by the executive directory." He made grim note of the irony that since the decree of October 1793 authorizing the transfer of the remains to the Pantheon, Condorcet, who had promoted the idea of Descartes as founding father of the Revolution, had himself been cut down by its violence. As Chénier's talk goes on it becomes clear that there is a rupture in the chamber over the pantheonization of Descartes, and to some extent lines are drawn with reference to how people view the Revolution.

The radicals, leaders of the Terror—"anarchistic tyrants," Chénier calls them—represent a deformity in the reason that underlay the Revolution, and these same people now wanted to deny modernity's forefather his rightful honor. "The persecutors of

Condorcet in life do not want to honor Descartes in death," Chénier charged. He reminded his colleagues again of the "numerous
services that Descartes rendered to humanity." He rolled out the
list of names of men who had contributed to the transformation
of knowledge of which they were the beneficiaries—Locke, Newton, Leibniz, Galileo, Kepler—and asserted Descartes' primary
place among them. He then tolled instances of the "ignominy of
the hereditary French government" toward their compatriot of a
century and a half earlier and concluded with a plea to carry out
the previous decree and transport the remains of this "great man"
to the Pantheon on the agreed-upon date.

Chénier apparently expected opposition, and it came in the
person of Louis-Sébastien Mercier, one of the most prolific and
opinionated writers of the day. At fifty-six, he had published volumes in virtually every literary form, but his greatest renown was
from two works that were themselves cutting-edge examples of
literary modernity. *Le tableau de Paris* and *Le nouveau Paris* were
guidebooks, compendiums of everything about the city and its
inhabitants that warned readers about stray animals and fog, advised on how to properly instruct a coachman, and digressed into
impressionistic observations of Paris at each hour of the day (at
two in the afternoon "those who have invitations to dine set out,
dressed in their best, powdered, adjusted, and walking on tiptoe
not to soil their stockings"). Mercier also wrote a bizarre proto-
science fiction novel called *L'an 2440*, a sensational best seller in
which he envisioned the city in that fantastically distant year.

When he was young and churning out prose by the bushel,
Mercier had composed a series of *éloges*—formulaic encomiums
to famous men, one of whom was Descartes. But he had since
changed his mind. He now rose and, evoking Chénier's florid delivery, began, "I, too, made an eloge to Descartes in my youth."

But he said that he hadn't yet realized that "the greatest charla-
tans in the world have sometimes been the men most celebrated."
Mercier chose to avoid combating Chénier's political argument.
Instead he railed against "the history of profound evil that Descartes has done to his country." Descartes, he declared, "visibly
retarded progress by the long tyranny of his errors: he is the father
of the most impertinent doctrine that has reigned in France. This
is Cartesianism, which kills experimental physics and which puts
pedants in our schools in place of naturalist observers."

Cartesianism, Mercier said, had taken root in schools and,
with its focus on theorizing rather than experimentation, had allowed the English to vault into the lead in science. But he insisted
his wasn't a nationalistic tirade. "We do not take offense at the
superiority of an Englishman," he said. "Newton belongs to all
humanity." But Descartes had led the French down the wrong
path in all the natural sciences—only in mathematics did Mercier allow that he had made a contribution. Mercier recounted the
previous burials of Descartes: in Stockholm, supervised by Queen
Christina, and in Paris, under the eyes of members of the church
and the Sorbonne. "I believe that these honors are sufficient for
the memory of Descartes and that his ghost has been entirely satisfied," he said. "The Pantheon is a republican temple that we reserve for the heroes and martyrs of the Revolution."

Mercier had a point—one that some would say is still valid.
The French have long had a cultural propensity to abstraction,
which they themselves have at times decried as counterproduc-
tive. Beyond national borders, there is that side of modernity that
prefers to ponder rather than act. Whole fields—sociology, literary and art criticism, history itself—have been accused of creating
self-perpetuating academic cults whose members talk exclusively
to one another without engaging the real world. Then there is

the irony that Mercier's diatribe against Cartesianism—its having rooted itself in schools and blocked progress—is precisely the charge that Descartes and his followers leveled against the Aristotelian system.

While there was truth in Mercier's criticism, it was also myopic. Another member stood up to express puzzlement. "Nature has ordered events so that the French Revolution came toward the end of the eighteenth century," he said. "However, I confess that, having heard this discourse, one has to ask oneself whether we are indeed moving toward the nineteenth century or if we find ourselves going backwards toward darkness." This member of the legislature saw the course of Descartes' life as following a pattern that had traced itself again and again since his time—more recently in the life of Jean-Jacques Rousseau, who had also traveled in many foreign countries as he fled criticism for his writings. Rousseau had been pantheonized two years earlier. "It will be enough to remember the career of Descartes to judge his genius and the homage to which he is due," the member said. "He was persecuted by kings and by priests; he was banished. . . . These persecutors, you find them persevering in pursuit of another celebrated writer of whom the memories are more recent. . . . The same men, I say, persecuted Descartes and Jean-Jacques." These were half-truths: Descartes had not exactly been persecuted by kings and priests, he hadn't been banished. But the desire to pinpoint historical precursors for the revolutionary struggle was irresistible.

The chamber was wavering now; apparently Mercier's points had resonated with some. Chénier took the floor again and addressed his colleagues in anger. "With regard to the project that I have presented in the name of the commission," he began, "I believe that the legislative body would cover itself in disgrace—" The rest of his sentence was drowned out by "violent murmurs," as the secretary noted in his minutes. "I cannot otherwise express my thoughts," Chénier continued after a moment. "I believe that the legislative body would compromise its glory and the national glory if, in ceding to the inclination that some persons seem carried away by, it denied today the solemn promise made to the memory of Descartes by the national convention."

The debate grew; Voltaire was brought into it, and members began to compare the revolutionary credentials of Voltaire, Descartes, and Rousseau. A member rose to clarify the differences between the contributions of Voltaire and Descartes. "Voltaire enlightened all classes of the people," he declared. "He employed, to be understood by each, the language suitable to them. The works of profound philosophy don't carry to all the world. As for Descartes, I have read part of his works, and I avow that never have I known so great a genius. I have read also Newton, but I have more veneration for Descartes, because he was first, and maybe also because he is French. I ask that the project of Chénier be instantly adopted."

There was too much disagreement, however. Someone suggested postponing the matter. Chénier said he would agree to a postponement, but he insisted that if the pantheonization of Descartes was attacked a second time, "I ask that nothing be decided until we have heard, at this platform, from all those who want to defend the Enlightenment and philosophy."

THE POSTPONEMENT PROVED FATAL to Chénier's cause. Years passed, and Descartes' bones stayed in the garden of Lenoir's museum. The museum itself continued successful, but much else changed. France's wars against the monarchies of Europe brought about the end of its revolution—not through losses to foreign powers but through the rise of one of the revolutionary government's own military commanders. From victories in Italy and Egypt, Napoleon Bonaparte returned in 1799 to conquer his own country, upending the weak government of the Directory and installing himself in power. The most notable change he enacted as "first consul" was to grant the Catholic Church some of its former status. Then, in 1804, having consolidated power, he changed his title. He was now emperor. The democratic republic—and, seemingly, the whole raft of dreams and ideals based on the reorientation of society around reason, science, and the individual—was finished.

If Napoleon represented a grand problem for France, and for Europe, he was a particular problem for Alexandre Lenoir. Lenoir now found himself in an awkward position. His was a "revolutionary museum," born out of the chaos of the Revolution, dedicated to the values of the Revolution. He tried to resell it to the new regime, pitching it no longer as offering a vision of history building to a climax with the Revolution but rather as celebrating the French past. He focused his efforts on Napoleon's wife, Josephine, and managed to get her to pay a visit to the museum along with members of her entourage. She came in the evening, and Lenoir had the building and gardens tricked out with flaming torches, the better to show off the sepulchral charms. Napoleon himself visited once as well and remarked that the exotic gloom—stony figures recumbent beneath a blue ceiling stippled with painted stars—reminded him of Syria.

To some extent Lenoir's effort worked: the museum endured through Napoleon's reign. But after Elba and Waterloo, with Napoleon's passing and the restoration of the Bourbon monarchy in 1814, Lenoir's luck ran out. Completing one of the most notable of the many pendulum swings between secularism and religion that have characterized the modern centuries, the Catholic Church came back into power alongside the monarchy, with renewed force. As one aspect of its return, individual churches around the country demanded to have their property back. Lenoir tried to appease churchmen and keep his collection together by proposing to add a religious focus to the museum. His idea was to group the tombs together in a chapel that he would design and to offer masses there. But it didn't play. In 1816, Louis XVIII issued a decree that religious property in the Museum of French Monuments be returned to its originating institutions; the same year, the museum's grounds were given to the national art institute. The Ecole des Beaux-Arts continues to occupy the site today.

Lenoir oversaw the dismantling of the collection he had personally amassed. Many objects went back to the churches from which they had been taken. Others went to the Louvre, where they remain to this day. The statues of French kings were returned to the Basilica of St.-Denis—and with them, too, went Lenoir. In recognition of his work in preserving so much of the nation's and the church's patrimony, he was given the position of conservator of monuments at the basilica. He spent the rest of his life there and continued to catalog art and artifacts until his death in 1839. His son Albert picked up where he left off, becoming a founder of the new field of architectural history and spending twenty-seven years compiling his massive three-volume *Statistique monumentale de Paris.*

When the Museum of French Monuments closed, the question of what to do with the tombs of so many French notables excited some popular interest, and various officials weighed in. One idea concerned the vast cemetery of Père-Lachaise. It had been organized under Napoleon but was so far from the city center it got little business. The idea involved transferring the famous remains from Lenoir's former establishment to the cemetery and making the occasion a public event in which the ancient, historic remains would give the new cemetery on the far eastern fringes of the city some attention and cachet. In March 1817, the city's conservator of monuments wrote to the minister of the interior and the prefect of the Seine proposing that the tombs of Descartes, Abelard and Heloise, the poet Nicolas Boileau, and the scholars Bernard de Montfaucon and Jean Mabillon be included, saying that "all these illustrious personages merit the same homage and the same religious treatment." The officials agreed, tombs were transported en masse, and the plan worked. The presence of the tombs of the great Molière, the poet La Fontaine, and especially the doomed lovers Abelard and Heloise—whose tragic love affair took place in the context of a twelfth-century version of the clash between faith and knowledge—excited morbid interest and encouraged cultured Parisians to buy plots. Today Père-Lachaise—whose more recent residents include Chopin, Oscar Wilde, Gertrude Stein, Edith Piaf, and Jim Morrison—is one of Paris's most popular tourist sites.

But Descartes wasn't part of the mass migration to Père-Lachaise. Once again, it seems, a group of "friends of philosophy" took a particular interest in his bones and exercised their influence. The cemetery was too remote. If the church of the patron saint of Paris had once housed the remains of the father of modern philosophy, another site, equally symbolic, had to be found. They settled on the church of St.-Germain-des-Prés, on the left bank of the river. It was the oldest church in Paris, founded in the sixth century, and the intertwining of its history with that of the city extended right to its partial destruction during the Revolution. On February 26, 1819, yet another formal religious ceremony—the third—was held over the remains. In the presence of the commissioner of police, the mayor of the Tenth Arrondissement, and delegates from the prefect of the Seine, the remains of Descartes, together with those of Mabillon and Montfaucon, were taken from the garden of the former museum. They were "extracted" from their tombs *"avec une religieuse attention"* and placed in fresh oak coffins. A party consisting of numerous members of the French Academy of Sciences processed with the coffins the short distance along the left bank from the former convent to the church. Here they were buried, and three black marble plaques erected, in a chapel on the right side of the nave.

The plaques can still be seen in the church today; Descartes' gives a fusty lineup of Latin platitudes extolling his immortal accomplishments. But exactly what lies beneath the plaque bearing his name is a matter of contention. When the porphyry box was opened, the members of the academy who peered into its ancient recesses were confused and dismayed at what they found—as well as at what they did not find. Something was wrong. Things were not as they had been led to believe.

Eventually the learned gentlemen set about doing what, as good, modern scientists, they were trained to do: analyze information, sound old theories, and construct new hypotheses. The bones of Descartes were about to leave history and enter science. Or, to reference another modern construct—a literary one—that was just being invented, they were about to become the subject of a detective story.

4

The Misplaced Head

Stockholm, April 6, 1821

Monsieur,

I have the honor of making to you a somewhat curious communication. In a meeting of your Academy of Sciences, where I was present during my sojourn in Paris, I heard the report made by members of the Academy who had been present at the transport of the bones of Descartes, I believe from the Church of Ste.-Geneviève to another place. It was announced that there were parts of the skeleton missing, and, if I am not mistaken, that the head was missing.

IF THE SAGA OF DESCARTES' BONES CAN SERVE AS a metaphor for modernity then it is doubly symbolic that during their peregrinations the head somehow got separated from the body and was to become, as it wound its way through the centuries, a source of mystery for various thinkers, artists, and scientists. For what does Descartes stand for today if not the cerebral over the material—the head over the body? Who bequeathed to us the mind-body problem?

In the seventeenth century it was considered normal for thinkers to cover a fairly astonishing range of subjects in depth—all of reality, more or less. A Descartes or Hobbes or Leibniz might devote one major work to light and optics, another to geology, another to God, another to free will, another to the movement of the tides, another to that of the planets. As the 1700s went by such grandeur became less and less possible. A botanical treatise written in 1542 listed five hundred known plant varieties. By the end of the 1600s the catalog included ten thousand; in 1824 the Swiss botanist Augustin Pyramus de Candolle indexed fifty thousand plants. Specialization was by this time the only way to advance knowledge. Descartes and other natural explorers of his day believed that nature was a puzzle that just the right sequence of discoveries would unravel, leading to astounding changes that they could not even imagine. Of course, they were right about the astounding changes, but they were naïve in their appreciation of the puzzle's complexity.

By the 1800s, there was much greater awareness of the complexity. The task of understanding the universe was in the process of being divided into different fields, and there was sometimes a geographic slant to the specialization. It so happened that the country in which Descartes had chosen to die was particularly rich in mineral deposits, making it a center of the newly emerging field of chemistry. Swedish chemists discovered a good portion of the sixty-eight elements known by the late nineteenth century, including oxygen, the elemental element, as it were (though since Carl Wilhelm Scheele's discovery of oxygen in 1773 was beaten out for publication by Joseph Priestley, credit is often given to both men).

The greatest of these chemists—and one of the most prominent figures in the history of science—was Jöns Jacob Berzelius,
the man who in 1821 sat down to compose the idiosyncratic letter
whose opening is quoted above. He started life in rural Sweden,
threshing hempseed and sleeping in the potato storage, launched
himself in a career in medicine, but discovered that he loved experimentation and analysis more than healing. He found work
at the School of Surgery in Stockholm under a revered professor
of medicine and pharmacology named Anders Sparrman, lived
in the house of a mine owner, and roomed with a physician who
operated a spa where people went for its healing mineral waters—surrounding himself, in other words, with chemicals and
chemistry. The young Berzelius was perennially short of cash and
made a deal whereby in lieu of paying for his meals he would create new mineral water mixtures—seltzers, bitters, alkalines, "liver
water"—for the spa goers' pleasure.

Under Sparrman, Berzelius did some things—discovering the
odd element—that today might merit a Nobel Prize, but when
Sparrman retired Berzelius was passed over as his replacement.
He was about to resign himself to a career as a country doctor
but the young man who had been given Sparrman's position died
suddenly and Berzelius was granted one of the few jobs in chemistry that then existed in the world. He was a bluff, florid man
with a capacity for enormous energy, and he went at his work
with herculean intensity. One of the hurdles the field had to clear
was determining the atomic weights of each element, which was
necessary in order to know how elements could be combined with
one another to form new compounds. In a feat of intellectual
and physical labor that has become legendary among chemists,
Berzelius set about fixing the weights of all the elements then
known. He typically worked from six-thirty in the morning to

ten at night; at one point he was nearly blinded in an explosion. The rewards were sweet. After fixing the combinations of silver chloride and sulfuric acid and barium hydroxide, he noted that "it is impossible to describe the bliss. . . . But to this end two years of ceaseless work had been given." He published his results in what quickly became the standard textbook of chemistry. Meanwhile, hampered and annoyed, as others had been, by the chaotic terminology and symbols that various scientists had devised for the elements and their combinations—some looked like hieroglyphs or a child's drawings—he invented what he thought was a clearer system, using letters from the beginning of the Latin terms for each element. He thus gave the periodic table—and the landscape of chemistry—the look that it has today.

Immediately following this burst of effort Berzelius suffered a nervous breakdown. Friends suggested travel as a way to recuperate, and he set out for the two capital cities of science. He was an international celebrity now and was received into the scientific inner circles in London and Paris. In London, this was the Royal Society; in Paris, it was the Academy of Sciences. These institutions reflected the different approach to science as it evolved in the two countries. The English were freelancers, and the Royal Society was something of a gentleman's club. But if the top-down approach of the French retarded the growth of industry in France, it had an important benefit for the development of Western history. As an offshoot of the government, the Academy of Sciences was able to function with an authority that the Royal Society did not. As Maurice Crosland, a historian of science at the University of Kent in Canterbury, notes, this authority in defining what science was started with the word itself. The Royal Society had an approach to knowledge that was holistic and at times playful, one that hearkened back to the seventeenth-century natural philosophers. In the early nineteenth century its members still tended to use the word *science* in the broad medieval sense, so that theology could still be regarded as "the queen of the sciences." Crosland argues that it was the French Revolution that nudged the members of the Academy of Sciences to begin restricting the use of the word to a particular type of secular investigation of the natural world, so that while in English *science* came into its current usage only in the 1830s, the Académie des sciences showed in its name that the French had long before moved in this regard in the direction of modernity.

The academy adopted an appropriately scientific approach to science, organizing itself into divisions, subdivisions, and sub-subdivisions, and in so doing helped define to this day the way knowledge is structured in university departments and research institutes. Astronomy, geography, chemistry, physics, mineralogy, botany, mechanics, agriculture—each branch had a department, and each department was tied to a school where that field was taught. Each held conferences, awarded prizes, funded research. When it was felt necessary, the members of the academy met to discuss whether to create a new subdivision, such as when the growing collections of fossils all over Europe led to the creation of a division of paleontology and then a subcategory of paleobotany.

Dating back to the period before the Revolution, the academy also defined what science was not. When Franz Mesmer came to Paris, having been run out of Vienna after causing an uproar with his "animal magnetism"—a precursor to hypnosis—members of the academy met in 1784 to consider whether "mesmerism," as it also became known, had any scientific basis. Mesmer's technique used magnets, long gazes, and pressure on the hands and arms to cause changes in patients; he argued that there was an unknown fluid, or "tide," within the human body that could be shifted in this way to bring about healing. All of Europe was in a frenzy over whether animal magnetism was real or not. The Faculty of Medicine and the Academy of Sciences decided to weigh in. The committee of review they put together was something of an all-star team of eighteenth-century science, including Lavoisier, the father of modern chemistry, Joseph-Ignace Guillotin, whose name was soon to become attached to the signature device of the Revolution, and, as visiting authority on electricity and other currents, Benjamin Franklin. In an early instance of the use of placebo and the single-blind trial, the scientists told some people they were being magnetized when in fact they were not, while others were magnetized without their knowing it. Those who had been told they were being magnetized, even when there was no actual magnet used, reported positive results; those who were magnetized without their knowing it showed no results. The scientists concluded that the evidence did not support the idea of the movement of tidal fluids within the body. Rather it demonstrated the effects of "the imagination." The academy deemed that mesmerism was not science—and from that moment it was not. Mesmer left Paris the next year. Mesmerism went on to have a lively career in nineteenth-century America, but it eventually went the way of the wooly mammoth and ultimately history would demote Franz Mesmer's immortal status from imposing noun to ephemeral adjective.

Berzelius arrived in Paris in 1818 as a guest of the academy. He was awed by Paris and the great houses where he was entertained and was fascinated both by the egalitarian nature of the salons ("In conversation there is no distinction between those of high station and other good folk. Titles such as Prince, Count etc. are never used in speech") and by how highly evolved his field had become in the city ("I believe that there are here more than

100 laboratories devoted to research, and there are several dealers specializing in chemical glassware whose stock is a source of astonishment to a poor Stockholmer who, when he needs a simple retort, cannot obtain it in less than three months"). Salons may have displayed a democratic sensibility but the academy itself, the inner sanctum of European science, impressed Berzelius with its grandeur. The members even wore a specially designed costume, green with frilly gold trim (they were French, after all), which amounted to a scientific uniform. Berzelius's trip was meant for recuperation, but he now felt rejuvenated enough to settle into a furious round of activity. With Claude-Louis Berthollet and Pierre-Louis Dulong, two of the great chemists of the day, he devised a way to further refine his calculation of the atomic weight of hydrogen. He met the discoverer of fatty acids and the discoverer of hydrogen peroxide, whom he admired (though the chemical had not yet evolved into a fashion statement), and he worked on a French translation of his book.

As it happened, Berzelius was in Paris when the third burial of Descartes' bones took place. One of those who had been invited to the burial ceremony was Jean-Baptiste-Joseph Delambre, the leading astronomer of the day and one of the two permanent secretaries of the academy, who functioned as codirectors. Delambre had a passion not only for science but for its history. His devotion to the ideals of precision and accuracy resulted in an achievement that has shaped the world to this day. Nearly thirty years before—in the midst of the Revolution—he had opened up a whole front of modernity by directing the project that led to the creation of the metric system. Throughout the centuries of the Middle Ages European localities had devised hundreds of different units of weight and measurement, which varied from village to village, and even those with the same name varied, so that a pound of bread or a pint of beer meant different quantities in different places. This nonsystem maintained local traditions but inhibited trade—on a very practical level it kept Europe medieval. The new, modern idea was to give everyone in the world one system, which would be based not on custom or legend or ancient myth but on nature—to be precise, on the scientific calculation of a natural standard.

The Revolution was a fitting milieu for such an idea to arise in, but it was also a dangerous one. A committee of the revolutionary government—the Commission on Weights and Measures—decided that the new base unit, the meter, should be related to the size of the globe, specifically that it should equal one ten-millionth of the meridian passing from the equator through Paris to the North Pole. Calculating this distance meant traversing a portion of it—basically the length of France—with sophisticated equipment for sighting and triangulating in order to obtain accurate measurements of individual stretches of the distance. This was the work that Delambre had undertaken as a younger man, and it was treacherous going. With a war on, he and his team of scientists and assistants—peering through scopes, adjusting sights, scribbling notes—appeared to the roving bands of revolutionaries and counterrevolutionaries as spies, and Delambre had dodged bullets and suffered imprisonment in fulfilling his mission. It would be some time before the metric system would be generally adopted (the first countries to accept it, the Netherlands and Belgium, would adopt the system just two years after the 1819 reburial of Descartes' bones, and France itself would not do so until 1840), but Delambre had long since achieved international fame for it in the scientific community in addition to his accomplishments in astronomy when he agreed to participate in what he thought would be a pro forma ceremony to rebury the bones of the founder of modernity.

The old astronomer, then, duly oversaw the removal of a wooden casket from the porphyry sarcophagus in which Lenoir had placed the remains of Descartes, then marched in procession the few blocks up the hill from the garden of Lenoir's former museum to the church, where, amid the stony medieval chill, the interior box was opened. What was found inside was remarkable enough that Delambre took notes and even included an account of the burial and contents of the coffin in his *History of Astronomy,* which he was just completing. "On an interior cask was attached a lead plaque," he wrote, "on which, having cleaned it off, we could read a very simple inscription, carrying the name of Descartes, the date of his birth and that of his death." Other than this, the officials were surprised to find only a few bones of recognizable shape; the rest was bone fragments and powder. The man who did the unpacking, Delambre added, "took some handfuls of powder to show us." The assembly then watched as these meager remains were placed in the vault that had been opened to receive them and were sealed behind a heavy stone.

Others who were about to become caught up in the puzzle of Descartes' bones would use words like "religious" and "precious relics" to describe the remains, but Delambre's interest was different. At seventy, he was an old-guard atheist from the heyday of the Revolution and the Enlightenment; he had no truck with religious or spiritual sentiment. His interest was in science and historical accuracy. The contents of the coffin that had been in Alexandre Lenoir's keeping seemed to tell a different story from the one that had been presented to Delambre and others. If the bones had been buried properly they would surely have survived the 169 years since Descartes' death in a better state of preservation. Were these Descartes' bones at all? Had he and the others solemnly buried the wrong remains? If they were in fact Descartes', how did they come to be in such a condition?

But while Delambre's interest was piqued, he didn't pursue the matter further but only went back to his home and jotted down what he had witnessed. He did, however, discuss these observations with some of his fellow scientists, either before or after a meeting of the academy. Several other of these men had also been at the reburial, and as they talked they came back again and again to the skull. A human skull would survive relatively intact even under somewhat adverse conditions. It seemed inconceivable that it would be reduced to powder. The only conclusion was that it had been separated from the rest of the body. Apparently one of the scientists had done some investigating, and he reported hearing a suspicion that the skull had never been among the remains in France—that it had never left Sweden.

Berzelius, the Swede, was party to this learned gossip. He expressed indignation. If, somehow, one of his countrymen had separated the skull of the great Descartes from the rest of the bones—Berzelius didn't shy from religious terminology and called it "certainly a precious relic"—then all Swedes should be reproached for such a "sacrilege."

And there the matter ended. What else was there to do but remark on the strange facts and then leave them to molder along with the remains? Delambre attended to his duties as permanent secretary of the academy. Berzelius—his period of recuperation over—returned home and took up new duties that mirrored those of Delambre, as the new secretary of the Swedish Academy of Science.

Two years passed. Then one day in March 1821, Berzelius

opened a Stockholm newspaper and found his attention caught by
an article on the estate of the late Professor Anders Sparrman—
the man whom Berzelius had first worked under at the School of
Surgery, and whose position he had eventually taken. "Something
curious has been noted recently," the article read. "At the auction
following the death of professor and medical doctor Sparrman,
the skull of the famous Cartesius was sold. It is said to have been
purchased for 17 or 18 riksdaler."

Berzelius was stunned by the coincidence—that he had been in
Paris when it was discovered that the skull of Descartes was missing and that, apparently, it had been in the possession of a man he
himself had known. He immediately went to work. He contacted
the auction house, and found that the skull had been bought by
a casino owner—and evidently a fairly infamous figure—named
Arngren. The vogue for maintaining a "cabinet of curiosities"—
bones, tusks, fossils, carved artifacts, feathered headdresses, seed
pods, fertility charms, butterflies, dried dung: a microcosmic attempt to make order out of the teeming chaos of the natural and
anthropological worlds—was then at its height, and Arngren
seemed to have thought the skull of the great thinker would make
a nice addition to the one on display at his casino. Berzelius went
to him, outlined the history of the bones of Descartes, and explained that it had recently been discovered in Paris that the head
was missing. To Berzelius's surprise, Arngren agreed to give the
object to Berzelius for what he had paid for it.

Berzelius then sat down to write the letter quoted at the start
of the chapter, which accompanied the remarkable object itself,
the skull of Descartes. His closest associate at the academy in
Paris was Berthollet, his fellow chemist, but he thought it best
to write to the biologist Georges Cuvier, both because Cuvier
had also taken a keen interest in Descartes' bones and because

he served alongside Delambre as the second permanent secretary. On receiving the skull, Cuvier decided it would be kept in the Museum of Comparative Anatomy, which was part of the Museum of Natural History. But he wasn't about to store it away just yet. It merited special attention.

A T LEAST AS MUCH as others who would be associated with Descartes' bones—Rohault, Condorcet, Alexandre Lenoir, Delambre, Berzelius himself—Georges Cuvier personified a major aspect of modernity. Indeed, all three of the men who involved themselves with the bones at this stage made their names in association with what was the principal scientific concern of the day: classification and measurement of the overwhelming amount of data that was coming in from all quarters. Delambre brought into being what would become the global standard of measurement. Berzelius developed the modern method of representing the chemical elements and ascertained how they combine to form virtually every substance on earth. The situation in biology was particularly complex. Biologists craved the sort of base principles that Newton had developed for physics. Trying to classify life-forms begged the question of what overall purpose you had in mind. The system that was still largely in effect in the early nineteenth century was the "teleological taxonomy" created by Aristotle and refined by the Scholastic philosophers: the "scale of beings" system, which the French called the *série,* or series, and which is popularly known as the "great chain of being." As with the medieval system of bodily humors, it was far more complex and useful than its popular stereotype suggests, but it had a serious limitation, which was its teleological basis. Teleology refers to an end

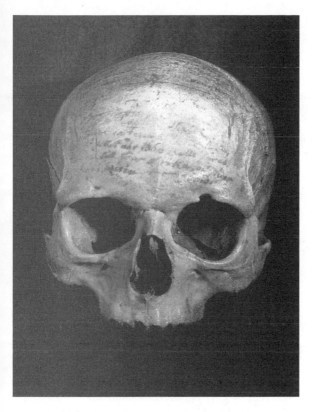

LEFT The skull of Descartes. Across the forehead, in Swedish, is an accusation of a theft in 1666 that began the skull's peregrinations. Above it is a poem in Latin celebrating Descartes' genius and mourning the scattering of his remains.

BELOW An eighteenth-century depiction of the court of Queen Christina in Stockholm by Pierre Louis Dumesnil. Descartes is standing to the right; the Queen is seated opposite him.

A seventeenth-century image showing The Dam, the central square in Amsterdam, shortly after the time Descartes lived there.

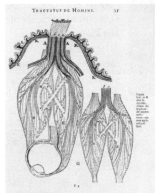

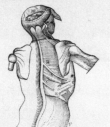

Anatomical drawings by Descartes illustrating mind and body interactions from his book *Tractatus de Homine*.

DISCOURS
DE LA METHODE /
Pour bien conduire fa raifon,& chercher
la verité dans les fciences.
PLUS
LA DIOPTRIQVE.
LES METEORES.
ET
LA GEOMETRIE.
Qui font des effais de cete METHODE.

A LEYDE
De l'Imprimerie de IAN MAIRE.
cIↄ Iↄ c XXXVII.
Auec Priuilege.

Title page of the first
edition of Descartes'
epoch-making work.

Drawings of Henricus Regius (left), Descartes' first "disciple," and Gysbert Voetius, who led
the attack on Descartes in the Netherlands in the 1630s.

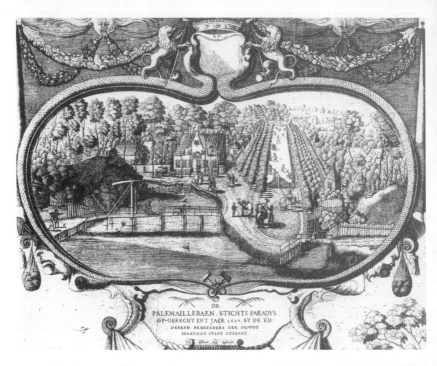

DE
PALEMAILLEBAEN, STICHTS PARADYS
OP-GERECHT INT JAER 1637. BY DE ED:
HEEREN REGERDERS DER OUWDT
BEROEMDE STADT UTRECHT.

ABOVE The location of Descartes' home in the Dutch city of Utrecht, as it was in the 1640s.

RIGHT The building in Paris that was the site of Jacques Rohault's weekly Cartesian salon in the mid-1600s. Today it houses a karaoke bar.

A meticulous eighteenth-century rendering of coffins and remains found in the old Church of St. Geneviève. It includes no indication of the coffin of Descartes.

The Church of St. Geneviève in Paris, on the right, was the second resting place of Descartes' bones, from which Alexandre Lenoir supposedly retrieved the remains during the upheaval of the Revolution. The church no longer exists.

Alexandre Lenoir, in a print showing him protecting tombs and monuments from the depredations of revolutionaries.

An eighteenth-century drawing depicting the interior of Alexandre Lenoir's Museum of French Monuments, which Napoleon said reminded him of Syria.

The decree of the revolutionary National Convention in Paris, dated October 1793, which gave Descartes, and his remains, special honors.

DÉCRETS

DE LA

CONVENTION NATIONALE,

Des 2 & 4 Octobre 1793, l'an second de la République Françoise, une & indivisible.

Qui accordent à René Descartes *les honneurs dûs aux grands Hommes, & ordonnent de transférer au Panthéon François son corps, & sa Statue faite par le célèbre* Pajou.

1°. Du 2 Octobre.

LA Convention Nationale, après avoir entendu le rapport de son comité d'instruction publique, décrète ce qui suit:

ARTICLE PREMIER.

René Descartes a mérité les honneurs dûs aux grands hommes.

II.

Le corps de ce philosophe sera transféré au Panthéon François.

III.

Sur le tombeau de *Descartes*, seront gravés ces mots:

Au nom du Peuple François,
La Convention nationale
à RENÉ DESCARTES.
1793, l'an second de la république.

The portrait of Descartes now credited to Frans Hals.

The bust of Descartes created by Paul Richer in 1913, with its breakaway face.

The classic portrait of Descartes in the Louvre, which was long thought to be by Frans Hals.

SKETCH IDENTIFIES SKULL OF DESCARTES

Drawing Made from Franz Hals's Portrait Removes Doubt of Relic's Authenticity.

CURIOUS HISTORY OF SKULL

Kept by a Swedish Officer, It Changed Hands Nine Times Before Reaching a Paris Museum.

Special Cable to THE NEW YORK TIMES.

PARIS, Jan. 25.—Whether a skull can be identified from a portrait is the interesting question raised by what is regarded as the final solution of the mystery of Descartes's skull. Prof. Paul Richer of the Academy of Fine Arts, to whom was intrusted the task of inquiring whether the skull in the possession of the Museum of Natural History was really that of Descartes, has compared it with Franz Hals's portrait of the great philosopher and now gives an affirmative answer.

Paul Richer's analysis of the skull of Descartes hit the *New York Times* on January 26, 1913.

The bust of Descartes, based on the skull, created for the 2000 Great Exhibition of the Face in Tokyo.

or ultimate purpose and typically means a religious purpose, as in God's plan. Aristotle's orientation of knowledge was teleological, which made it easy for the Scholastics to adapt it to conform with a Christian view of creation, so that as the chain of life-forms proceeded from the simplest organisms to more complex ones it also reflected a spiritual hierarchy. In the eighteenth century this system began to lose its usefulness. By the early nineteenth century biologists and botanists had a frank case of "physics envy"— a yearning for a Newton-like figure to create underlying laws to ground their science. Cuvier argued for a completely new system that ignored teleology and instead grounded itself in observation of bodily parts and their functions. In coming up with his system, he helped invent modern zoology and comparative anatomy.

Cuvier's work built on that of his predecessor, the Swede Carl Linnaeus, who had ranked living creatures into kingdom, class, order, genus, and species. Linnaeus had looked to the parts of the reproductive system as a basis for classifying and differentiating creatures, but while reproduction was undeniably elemental, it was not necessarily the most useful organizing principle. Cuvier based his system instead on the correlation of parts and how those parts worked together and in a creature's environment (an animal with sharp claws also tended to have teeth appropriate for tearing into the prey it captured). He divided animals into four categories based on body structure—vertebrates, mollusks, articulates (for example, insects), and radiates (for example, starfish)—a system that remained basic to biology for most of the modern era. Cuvier applied his system with an almost mathematical logic. A ruminant animal, by definition, had to have a forestomach in which it partially digested its food, so if you found an animal with such an internal arrangement you knew it was a ruminant, and conversely if you found an animal without a forestomach it could not possibly be one that digested its food in two stages. A perhaps apocryphal story has some of Cuvier's students wrapping one of their number in a cowhide and challenging their teacher to identify the beast. As the master entered the room, the student cried, "Cuvier, I am the devil, I've come to eat you!" Whereupon Cuvier is supposed to have replied something to the effect of "Don't be ridiculous. You have a divided hoof, therefore you eat grain."

Comparative anatomy was only one of the fields that Cuvier pioneered. His interest in bones led directly to the study of fossils. Comparing skeletons of extinct animals of different geological eras led him to conclude that the earth had endured several prehistoric cataclysms, resulting in mass extinctions. He noted, of course, the slight alterations in skeletons of seemingly related creatures but did not argue in favor of a theory of evolution. It was in the air at the time; his countryman and colleague at the academy Jean-Baptiste Lamarck had recently advanced an evolution argument. But, to the contrary, despite his many groundbreaking accomplishments, Cuvier is best known today for retarding the advance of evolution in the scientific community, keeping it from serious consideration until the publication of *On the Origin of Species* in 1859. Technically, scientifically, Cuvier's objection was based on his theory of the correlation of parts. Nature, he wrote, "has realized all those combinations which are not repugnant and it is these repugnancies, these incompatibilities, this impossibility of the coexistence of one modification with another which establish between the diverse groups of organisms those separations, those gaps, which mark their necessary limits and which create the natural *embranchements,* classes, orders, and families." That is, every species that exists, or that ever did exist, was a functional whole that needed all of its parts to be just the way they were. A slight mutation in one part would collapse the whole system.

Evolution based on small changes over generations was thus impossible. Instead, Cuvier argued for the opposite—"fixity of the species"—and did everything in his power as permanent secretary of the Academy of Sciences to advance that view.

There was a less scientific aspect to Cuvier's views on evolution. He was a devout Christian, and the early nineteenth century was a time when Christians were using science to undergird the Bible. Such efforts, of course, went back to Descartes himself, who believed that his mechanistic theory of nature was in fact a defense of Christianity—that it "bracketed" the material world, making it the domain of science and leaving theology free to treat the human soul. Belief in science had grown enough by the early nineteenth century that even quite literal-minded Christians tended to look to science for evidence to support, for example, the biblical account of creation or the flood that Noah navigated.

Cuvier was a rigorous scientist and he didn't overtly manipulate data to support Christian accounts. However, he was interested in showing that science and faith were compatible, and his approach to biology and paleontology reflected that. The problem was that scientific evidence seeming to contradict basic parts of the Bible had become mountainous. He took account of the same evidence that Darwin would use to argue for changes in species over time based on natural selection but made it serve the opposite theory, one that squared with biblical views of creation. In a way, his argument has a curiously modern sound. He believed in God unquestioningly—in fact regarded it as a misuse of reason to question God—and his belief underlay his science, including his views about the "repugnance" that nature had for changes to its design. For Cuvier, in other words, nature showed the intelligence of the Creator, and the idea of species evolving over time, buffeted by random forces, was repellent both to this larger intelligence and to human intelligence. Beneath his science, then, is a nineteenth-century variant of the very current "intelligent design" theory put forth by Christian thinkers who, like Cuvier, believe that the theory of evolution does harm to the Christian account of the world.

Cuvier was nevertheless one of the models of a nineteenth-century scientist, and as such he had a love of his field's development, its increasing complexity, and also its beginnings. When, in May 1821, he received, from the hands of the Swedish ambassador to France, a package sent by Berzelius from Stockholm along with the letter describing his serendipitous discovery, he opened it with something approaching awe. Like Delambre, he had been dismayed by the discovery of the sorry state of the remains during the reburial two years before and by the absence of a skull. Now here was an object that seemed to deepen that mystery and to open another.

To be sure, the skull was no ordinary object—it wasn't even an ordinary skull. In the world of art and old paintings, provenance—a paper trail, authenticating proof of a chain of past ownership—is everything. This skull seemingly arrived with its own provenance, which deepened Cuvier's interest. He contacted Delambre at once, and the two put the subject on the agenda at the academy.

On April 30, 1821, the Academy of Sciences met at its home on the banks of the Seine. The members included some of the most famous names in the history of science, among them Berthollet, who helped create the language of modern chemistry; Jean-Baptiste Lamarck, he of the evolution theory; Joseph-Louis Gay-Lussac, who formulated several laws of physics, codiscovered the actual chemical composition of water, and did the underlying work that led to the "alcohol by volume" calculation that is found on every bottle of wine, beer, and spirits; Pierre-Simon Laplace, who extended the work of Newton in mathematical physics and theorized on the origin of the solar system; and André-Marie Ampère, a discoverer of electromagnetism, for whom one of the basic units of electrical measurement is named. They heard a report on the inflammation of the membranes of the central nervous system. A member named Poyet presented some examples of new methods of bridge construction he had developed. There was a report on the medicinal properties of flowers of the Antilles. Then the assembled luminaries gathered around the object that Cuvier placed before them and studied it with, as the chemist Berthollet said, "a religious reverence." Cuvier read Berzelius's letter recounting how he came to be in Paris at the time Descartes was reburied and heard of the absence of a skull among the remains and how, only a month earlier, he discovered that a skull purported to be Descartes' had come up for auction. "Our minister in Paris, M. le comte de Löwenhjelm, who left here the day before yesterday, was kind enough to take charge of the transport of this relic," Berzelius's letter said, "of which I pray you, Monsieur, to make use that you judge reasonable."

The skull was missing its lower jaw but was otherwise intact. With its empty black sockets it gazed back at this historic collection of wise men as if chiding them for whatever smugness they may have accumulated in their quest to advance knowledge, forcing them to ponder the hard limit they all faced, the remorseless indifference of death.

At the same time, it offered a challenge. For these were all men who had devoted their lives to solving nature's puzzles, for whom *method* had become second nature, and here was a puzzle about the man who was arguably the father of all their varied disciplines, the very originator of "the method." The tantalizing thing was that the skull was covered with intricate pen marks. Many were signatures, marks of ownership. But splayed across the top, in flowing Latin cursive, was a poem that fairly shouted at the observers:

> This small skull once belonged to the great Cartesius,
> The rest of his remains are hidden far away in the
>   land of France;
> But all around the circle of the globe his genius is
>   praised,
> And his spirit still rejoices in the sphere of heaven.*

The questions piled up. Who had written this, and when? What did "hidden" mean? Was it possible that the bones they had reburied in St.-Germain-des-Prés were not Descartes'? Was this indeed Descartes' skull, and if so how had it gotten separated from the body? What exactly had happened to the remains of René Descartes in the 171 years since his death?

A further clue—seemingly a fairly massive one—also offered itself on the skull. This one was right in front, scrawled across the forehead. It was written in Swedish, but Berzelius had provided a translation:

> The skull of Descartes, taken by J. Fr. Planström, the year 1666,
> at the time when the body was being returned to France.

---

*The poem and the signatures were mostly visible in Delambre's day but have faded considerably since. The poem reads in Latin:

> Parvula Cartesii fuit haec calvaria magni,
> exuvias reliquas gallica busta tegunt;
> sed laus ingenii toto diffunditur orbe,
> mistaque coelicolis mens pia semper ovat.

Cuvier had done some investigative work and was inclined to support the authenticity of the skull. For corroboration, he placed alongside it an engraved portrait of Descartes and pointed out for the savants what he took to be similarities in cranial features. But more work needed to be done. There were all sorts of oddities; for one, above the sentence about the mysterious Planström were some barely legible words—a name that was hard to make out, "1666" again, and once more the Swedish *"tagen,"* which Berzelius told them meant "taken." Cuvier wanted further information about "this precious relic." The members agreed that someone should continue the research. Then they moved on to other business, perhaps no less of interest, with a Monsieur Virey rising to present his paper on "The Membrane of the Hymen."

IT WAS DELAMBRE who took up the matter of investigating the skull. Delambre revered Descartes as a father of science. Delambre also knew that his own major work in the service of science was behind him. He was seventy-two years old and not in good health; the odd little task he was about to undertake could become a coda to his scientific career.

At the meeting of the academy on May 14, 1821, he presented his findings in a report that ran to three thousand words; his reading of it took up nearly the entire session. He titled it "Skull coming from Sweden said to be that of Descartes: Facts and Reflections," and indeed it was organized into a series of "Facts" each followed by a section labeled "Remarks."

But if his colleague Cuvier was anticipating an exhaustive confirmation of his own speculations, he was to be disappointed. Having completed his investigation, Delambre felt strongly enough

about the matter that he deemed it necessary to assume the role
of opposing attorney. Across the top of his report he scrawled his
conclusion: "M. Cuvier . . . believes that the skull is that of Descartes, because he finds great conformities with the engraving,
and I believe the opposite."

He began reading aloud. He gave his colleagues a history of the
pertinent events associated with the bones of Descartes, culmi-
nating with the third burial ceremony, two years earlier, at which
he and others were shown the contents of the coffin, whose mea-
gerness he characterized as "really a bit remarkable."

Then he turned his attention to the object before them. The
marks tattooing its surface were intriguing, he admitted. "But,"
he said, "what proof have we from elsewhere regarding its authenticity? Some inscriptions, more or less effaced, that one makes out
on the convexity, which are the names of the successive owners,
with some dates and nothing more." True, there was what seemed
to be a testimonial of some sort on the skull. But who was to say
who this Planström was? And what information could be had
about the presumed accuser who had written this barely legible
sentence about him? One could infer almost anything from it.
Even if Planström had indeed taken the skull in 1666, it didn't indicate when the skull was actually separated from the body. That
could have been done "either at the home of Ambassador Chanut,
immediately after the death, or in the provisional grave of 1650,
or in the tomb of stone, or in the presence of Terlon in 1666, or
finally at Peronne when the cask was opened by the customs officials." It was even possible, Delambre added, that it had been
removed for a particular purpose that was part of the historical
record. It was known that Chanut had had a death mask made
of Descartes, from which a French artist named Valary who had
been a regular of Christina's court had sculpted a bust (both the

death mask and the bust subsequently went missing). Could it not have been that the sculptor "separated the head from the trunk in order to cast it at his ease and that he then neglected to return it?" Delambre asked his colleagues. "One at least has to admit that it has some plausibility." That would mean there was a gap of sixteen years from the time the head was separated from the body until this novel record on the forehead of the skull indicating that Planström had "taken" it. How did one know that someone early on, knowing of the fuss over the unburying of Descartes' bones, had not had the notion to adorn a random skull with writing in order to perpetrate a prank or deception, perhaps to make money? And once the first owner was deceived, all the others would have accepted the provenance as genuine. All they had, Delambre insisted, was "one assertion lacking proof" that the brain inside this skull had once thought "I think, therefore I am."

All in all, Delambre considered it most likely that the skull had never left the body. The remains Delambre had seen in the coffin were mostly fragments, which, on consideration, suggested to him that the body had suffered severe exposure, and it was logical to suppose the skull had been treated similarly and had been reduced to a similar state: "It is not impossible that it is very well the same with the skull after 169 years." Delambre concluded with a request that his report be included in the register of the academy "in order that someone may respond to my objections or clarify my doubts."

It didn't take long for a response. Cuvier seems to have sat listening to his illustrious colleague with mounting bewilderment. Delambre's report was a stew of contradictions and irrelevancies. Cuvier rose to address his colleagues. In any analysis, Cuvier believed, common sense had a part to play in one's reflections. Why imagine, for instance, that the head might have been sawed from the body immediately after death in order for the bust to be created? What possible basis was there for so grisly a supposition? Cuvier chided his fellow secretary for manufacturing elaborate scenarios and suspicions. The "principle of parsimony," a tenet of science, says that it is better to prefer simpler explanations to more convoluted ones. Surely it was more sensible to suppose that this Planström had taken the skull than to imagine someone *not* taking Descartes' skull but using another man's head and inventing a fiction around it. Especially considering what had launched this investigation: the fact that when the coffin was opened two years earlier there was no skull found with the remains.

Besides, Cuvier asserted, the skull itself indicated when it had been taken from the rest of the bones: in 1666. That meant the year the remains were unearthed in Stockholm, at or around the time of the ceremony in the residence of the French ambassador. It seemed clear to Cuvier that "the moment when this head would have been removed must be when the bones were packed up to be taken to France." As to who took it, there was clear evidence to that as well—not conclusive but compelling: "Planström." Find that man, and the mystery would begin to unravel.

Cuvier might have wondered about his colleague. Physically, Delambre, as he laid out his views on Descartes' skull, probably presented to his fellow scientists a reduced figure from what he had been even a few years before. He was ill enough that he was making arrangements for his own death (he would be dead the following year). Could senility or some weakening ailment have affected his reasoning powers? Every member of the academy knew of Delambre's tremendous tenacity and determination, which either stemmed from or was exemplified by a childhood incident. When very young, he had been stricken by smallpox, which partially blinded him and, to boot, left him without eyelashes, giving him, for the rest of his life, a naked, preternaturally defenseless aspect, simultaneously intense and wan. Imperfect eyesight had led him to overcompensate: he read voraciously in his youth, willing his vision to improve, and it did. Then, like someone struck by a muscular disorder who vows to become a professional athlete, he developed a desire to peer into the tiny aperture of a telescope, to squint at points of light hundreds of thousands of miles away and see new things in them. He subdued his distinct handicap to the point of becoming the leading astronomer of the day, not to mention devising the standard of measurement that would revolutionize science.

But were his powers still intact? Cuvier must have wondered. Delambre seemed to have gone to great lengths to create difficulties in his analysis of the skull. It might be worth noting another possible explanation for Delambre's refusal to allow informed speculation into his thinking. He had a particular and highly unusual reason for sensitivity toward anything that smacked of uncertainty in scientific investigation—one that neither Cuvier nor any of his other colleagues in the academy could have known and that, if it had been known, would have caused a scandal.

Thirty years before, when Delambre led the team that did the calculations from which to derive the length of the meter relative to the circumference of the earth, by measuring the arc from Dunkirk to Barcelona, he had had a partner named Pierre-François-André Méchain. The two astronomers had divided the work, Delambre starting in Paris and going north and Méchain heading south. Late in the game, after years of toil, thousands of miles of travel, and untold calculations, and with the final result established, Delambre had made a devastating discovery. His partner had miscalculated, then covered up the error in his work. Delambre had made this discovery in 1810, eleven years before his encounter with Descartes' skull. He then made what Ken Alder, author of *The Measure of All Things* (2002), considers a fateful personal and professional decision. He covered up the cover-up. In the archives of the Observatory of Paris Alder found Delambre's handwritten notes, which had apparently lain unseen since he wrote them: "I have not told the public what it does not need to know. I have suppressed all those details which might diminish its confidence in such an important mission. . . . I have carefully silenced anything which might alter in the least the good reputation which Monsieur Méchain rightly enjoyed." Yet Delambre had preserved the logbooks with their errors, along with the acknowledgment of his own discovery. He had anguished over the issues of truth, error, and certainty and had made an allowance for error and uncertainty. Now, at the end of his life, working on a project of little real-world significance but a certain symbolic importance, Delambre was struck by a need either to have absolute certainty or else to dismiss entirely the skull's legitimacy. There's no way of connecting this need to the discovery of a scientific error and his decision to hide it, but Alder's own thesis is that Méchain's error and Delambre's discovery of it are important in the history of science because these twin facts show the dawning realization on the part of "hard" scientists that certainty is impossible, that error and inaccuracy are a component of their work. If Delambre's willingness to bury his partner's error indicates an awareness of the very modern notions of probability and error as facts of life, then his refusal to consider circumstantial evidence in support of the authenticity of Descartes' skull was perhaps a tug in the other direction—a reflexive urge to eradicate uncertainty, to purify.

Cuvier, meanwhile, thought the evidence regarding the skull strong enough that he took over the investigation. More needed to be known about the circumstances of the disinterment in Stockholm. He contacted a man named Alexandre-Maurice Blanc de Lanautte, comte d'Hauterive, who held the government position of archivist in the office of foreign affairs. Hauterive had had an exotic career, serving in Constantinople as part of the French diplomatic mission to the Ottoman Empire and in New York as consul to the newly founded United States before settling into a research position. Cuvier told him the particulars of Descartes' life and death and the names of the officials who had been involved in the transport of the remains. Perhaps an answer would be found in the official correspondence of the time.

Cuvier, Delambre, and Hauterive had access to the same early source on Descartes' life that we have today: the biography by the seventeenth-century priest Adrien Baillet. Cuvier noted that Baillet indicates in his biography that he had available to him letters from Terlon to d'Alibert, the treasurer of France, and a written note by Simon Arnaud, the marquis de Pomponne, who was in the process of succeeding Terlon as ambassador to Sweden and who attended the 1666 exhumation and the ceremony at the ambassador's residence.

Hauterive, however, was unable to find any relevant documents in the government files. "The correspondence of M. le Chevalier de Terlon, minister from France to Sweden in the years 1666 and 1667, makes not a single mention of the transport to France of the remains of Descartes," Hauterive reported to Cuvier.

But he did find something. He went on to add that "one had the idea to consult various printed works, one of which presented curious information on the object in question." Hauterive had discovered a Swedish work from the mid-1700s citing a chain of ownership of the purported skull of Descartes. There was an originating name and a story to go with it. The man who had taken the skull, according to this source, had played a role in the events surrounding the first disinterment. The source gave this man's name: "Is. Planström."

AND SO WE GO BACK to the beginning, which is to say the end—the dead of night in the dead of winter in the year 1650. In an upper-floor room of a building in central Stockholm, an ailing man breathes his last. After some dispute, it is decided to bury him in a forlorn cemetery a mile away, on the outskirts of the city. The suns of sixteen summers warm the earth in which the remains lie, and the deep cold of sixteen winters freezes it. Within the first 128 days—according to estimates of modern forensic anthropology and based on an average temperature of 50°F—the soft tissue would have decomposed. The bones would likely have begun to bleach in the first year. Before ten years had passed the bones would have started to exfoliate and crack. If the coffin was weak (it was described as "porous"), roots, rodents, and worms would hasten the decay.

Sixteen years after burial, then, the remains are dug up and hauled back to the same building in which the man had spent his final hours. There, in the chapel of the French ambassador's residence, a ceremony is held, presided over by officials of the Catholic Church in Sweden. The bones—which have become separated—are transferred to a copper coffin that is two and a half feet long. Hugues de Terlon, knight of St. John and French ambassador to Sweden, who is in the process of assuming his new post in Denmark, asks permission to take, as a personal relic, a bone of the right index finger. The coffin is kept, until the time of departure for Paris by way of Copenhagen, at the residence of

Terlon, where it is watched over by members of the Stockholm city guard.

Isaak Planström was the captain of this guard contingent. The details of his involvement came via Hauterive's information. Hauterive discovered that in 1750 a school headmaster named Sven Hof, from the town of Skara, wrote of traveling some years earlier to Stockholm to visit a friend and colleague named Jonas Olofsson Bång, who proudly showed him the skull of René Descartes. With the object came a story. The skull had come down to Bång from his father, a brewer and merchant named Olof Bång who had told him how he came by it. A man who had owed him money had died, and the elder Bång collected some of his property in lieu of payment. Among the items was the skull of Descartes. Bång told his son that the deceased, Planström, had had the job of watching over the remains of Descartes before they were to be shipped to France. Planström had explained his act by saying he felt that Sweden should not "lose completely the remains of such a famous person." Bång said that the guardsman kept the skull for the rest of his life as "a rare relic of a philosophical saint." Bång in turn kept it for the rest of his life and passed it to his son.

The younger Bång told this story to his friend Hof, then apparently said something to the effect that he would like to find appropriate words with which to adorn the object, whereupon Hof wrote out a few commemorative verses in Latin, and Bång later inscribed them onto the top of the skull. Hof's account included the Latin lines, which were identical to those on the skull that was now in the possession of the academy.

Berzelius, in his letter to Cuvier, noted the various names adorning the skull, some of which were illegible or only partly legible, and suggested that it should be possible to work out from them a history of the skull's career in Sweden. In the 1860s and 1870s, a man named Peter Liljewalch did that. Liljewalch was born in the Swedish city of Lund, became a medical doctor and specialist in infectious diseases, joined the army and traveled to Denmark, Germany, and Russia, and eventually became court physician to Desideria, queen of Sweden and Norway from 1829 to 1860. Sometime thereafter, Liljewalch returned to Lund and settled into the curious retirement project of charting the Swedish owners of the skull of René Descartes.

In the summer of 2006 I traveled to the manuscript division of the Lund University library, where a librarian laid before me the collection marked "Liljewalch." I opened folders filled with delicate sheets covered with notes in an elegant nineteenth-century hand. Liljewalch had followed trails backward in time, pursuing the lives and deeds of the men who, for one reason or another, had come into possession of the skull. Somehow the cranium made its way from the younger Bång to a military man named Johan Axel Hägerflycht, who kept it until he died in 1740. When his goods were dispersed the skull fell into the hands of a government official named Anders Anton Stiernman. His is one of the names still visible on the skull, on the right side, along with a year, 1751. When Stiernman died his son-in-law, Olof Celsius, found himself its owner and promptly placed his own signature on the occipital bone, at the lower back. This Celsius was a man of the cloth who became the bishop of Lund. His interest in the head of Descartes must have been as a scientific charm or talisman, for science was in the family. His father was a botanist, other elder relatives were astronomers or mathematicians, and his cousin Anders Celsius was the astronomer for whom the temperature scale is named.

At first blush Johan Fischerström, the next owner, an "economic superintendent" in Stockholm, does not seem to have satisfied the skull's penchant for finding its way to people who had pointed associations with one or another of the main features of modernity. But Fischerström was evidently a man of passion, and the notable aspect of his life turns out to have been not his career but his love interests. He became the object of affection of Hedvig Charlotta Nordenflycht, who is sometimes referred to as Sweden's first feminist. She was a serious student of philosophy and the doyenne of Stockholm's leading Enlightenment-era literary society, the excellently named Order of the Mind Builders. Nordenflycht became renowned for her mix of philosophical moodiness—over issues such as the role of reason in shaping moral decisions—and lovesickness. Her fame came from her poetry, in which she explored themes of nature and loss. She met Fischerström at a Mind Builders gathering sometime in the early 1760s, when she was in her early forties and he was in his late twenties, and fell for him. He left her for a younger friend of hers, and legend has it that she drowned herself as a consequence. The cad Fischerström lived on and procured the skull of Descartes sometime after to add it to his cabinet of curiosities.

Fischerström kept the skull until he died in 1796, and when his property was auctioned off it was bought by a tax assessor named Ahlgren, whose signature can today barely be made out behind where the left ear would be.

In the 1760s, while Hedvig Charlotta Nordenflycht was suffering the inattention of Fischerström in Stockholm, another Swedish devotee of eighteenth-century philosophy, named Anders Sparrman, was getting the best possible scientific education, under the tutelage of the great botanist Carl Linnaeus. On completion of his study Sparrman signed on as a ship's surgeon in order to travel to Asia. He returned after two years in China laden with samples of the country's flora and fauna, which formed the start of his own cabinet of curiosities. In 1772, fueled by a passion for collecting specimens of the natural world, he ventured to Africa, where he supported his collecting habit by dispensing medical advice and teaching the children of an official of the Cape Colony. One day late in the year a German named Johann Forster showed up, off a ship that was anchored in Table Bay. Forster was himself a naturalist, and the two instantly became fast friends. Forster avowed that there was no better place for an enthusiastic young naturalist such as Sparrman than on board the ship on which Forster was at that moment serving. Its captain, James Cook, was then in the midst of his second voyage of discovery, and Forster convinced Cook that the Swedish naturalist would be a valuable addition.

Having thus hitchhiked his way into history, Sparrman spent the next three years at sea with Cook, skirting the pack ice of Antarctica and rounding the Antarctic Circle, circumnavigating New Zealand and exploring Tahiti and other islands of the South Pacific. One part of Cook's mission was a modern challenge to received wisdom much as William Harvey's theory of the circulation of the blood was a challenge to the older system of bodily humors—or indeed as Descartes' philosophy represented a confrontation with the medieval system of knowledge developed by the Scholastic philosophers. The Royal Society had charged Cook with finding the continent of Terra Australis, which had been theorized by Aristotle and later writers to lie at the southern pole. The ancient thinking was that a giant land mass must exist at the bottom of the earth to counter those at the northerly extremes. Many members of the Royal Society remained convinced of the logic of this, though Cook himself had all but concluded, based on his first voyage, that it was not so.

Besides assisting in the voyage that would disprove the existence of Terra Australis—and that would in several other ways broaden knowledge of the world—Sparrman took copious notes and from them wrote *A Voyage to the Cape of Good Hope,* which became one of the classics of eighteenth-century naturalism and a standard source on Cook's career. Later, after mounting an inland expedition of his own into South Africa, Sparrman returned home to Sweden laden with specimens. He then made a trip to London to visit what were widely considered the greatest cabinets of natural objects.

By the time he settled down again in Stockholm he had been showered with honors as one of Sweden's great scientists. He took up a position as professor at the School of Surgery in the 1790s. In 1802, the young Berzelius won a position as his unpaid assistant. Three years later, Sparrman retired, and Berzelius eventually assumed his professorship. About that same time, Ahlgren, the tax official, apparently communicated to his friend Sparrman that he had come into possession of an object that he thought might interest the professor. The poignancy could not have been lost on Sparrman. He had traveled the globe gathering skulls and femurs, fibulas and fossils from every variety of creature. The skull that had given birth to modern philosophy would crown his collection.

So it was that while Alexandre Lenoir was tending to what were purportedly the rest of the bodily remains of Descartes in the garden of his Museum of French Monuments in Paris, the skull was in Stockholm in the collection of Berzelius's former mentor. Berzelius, who happened to be in Paris when the remains were removed for burial, thus became the link that reconnected mind and body, so to speak.

Or was he? Could the savants at the French Academy of Sciences conclude that the skull was authentic? None of them knew much of this history of the skull's life in Sweden, but the information from Hof supported the case. Cuvier and Delambre held a follow-up meeting at the academy five months after the first viewing of the skull, and Delambre presented a supplemental report in which he commented on the new information provided by Hauterive. It had its compelling features, Delambre admitted, but he remained skeptical, and he had some justifications for skepticism, for new difficulties came with the evidence of headmaster Hof. Hof asserted that "Isaac Planström, officer of the guard for the city of Stockholm, removed the skull from the bier of Descartes, *for which he substituted another.*" If that was true, it would explain why those who dealt with the remains later—the customs officials who opened the coffin at the French border in 1666, among others—did not report anything amiss. But it added a new mystery: what became of this second skull, the ersatz head of Descartes?

While the French scientists were puzzling over these questions, Berzelius happened to write to a Swedish friend telling him of the strange circumstances surrounding the skull of René Descartes and what he believed to be his satisfactory resolution of them. The reply he got may have incited a groan. The story sounded quite fascinating, wrote Hans Gabriel Trolle-Wachtmeister, a nobleman and government official with an amateur interest in chemistry, "but are you completely sure it's the right skull?" In Lund, Trolle-Wachtmeister informed Berzelius, was *another* skull of Descartes, "to whose authenticity the rector and council are prepared to swear." He added drily that in the scheme of things it wouldn't be unjust if it turned out the great Descartes had had two heads since "we see how many fools have one."

Was this, then, the second skull? Were there indeed two skulls of Descartes in circulation, one genuine, the other a placeholder slipped by Planström into the copper coffin? But if so, how was it that both had remained in Sweden? Wouldn't one of them have ventured to Paris with Terlon's party?

The complications were only beginning. In Paris, the savants were puzzling over another item. The information about Sven Hof was embedded in a four-volume biography of Queen Christina by a man named Johan Arckenholtz; it was in the first volume, which was published in 1751, that he included Hof's account of the encounter with Descartes' skull. By volume 4, which appeared in 1760, Arckenholtz had himself entered this exotic subplot: he reported that in 1754 he had "made the acquisition of a part of this skull that is attested to be genuine and of which the other part rests in the cabinet of the late M. Hägerflycht." So it seemed that there were now *four* skulls or skull pieces that had supposedly once belonged to Descartes. The situation was beginning to mirror the relic trade of early Christianity, when saints' bones proliferated and, as John Calvin wrote with Protestant scorn, there were enough pieces of the "true cross" circulating around Europe to fill a ship's cargo.

It's possible, though, to wade through at least some of the cranial debris. Trolle-Wachtmeister was correct about Lund University's possessing an object its keepers took to be a piece of Descartes. This object was probably what spurred Liljewalch to track the skull's owners in the 1860s. In fact, the object is still there. In common with many other historical museums around Europe, the Historiska Museet in Lund is built around a long-ago bequest of somebody's cabinet of curiosities. The museum has its touches of twenty-first-century museum studies—some breezy labeling and interactivity—but mainly it preserves its reassuringly old-fashioned origins. Kilian Stobaeus, a scientist who left his collection to Lund University in 1735 and in so doing founded what would become Europe's first archaeological museum, had a particular fancy for tribal objects from all over the globe. It's pleasantly jarring to encounter a sweeping collection of American Indian artifacts—arrows, baskets, jewelry, whole birch canoes— in the middle of Sweden. Hampus Cinthio, the historian on the staff who guided me through the collection, told me that American museums lust after these objects dating from decades before the American colonies fought their revolution, most of which are in an excellent state of preservation.

In the same room as the American Indian collection is a glass case that contains, among many other items, a portion of a human skull labeled in an antique hand "Cartesi döskalla 1691 N.6." In a juxtaposition surely not coincidental, it sits beside a pair of embroidered purple slippers, so small they look like they would fit a doll, that once covered the feet of Queen Christina. The curved piece of bone, about the size of a cupped hand, is the object to which Trolle-Wachtmeister referred. It is also what Arckenholtz had referred to sixty years earlier as the "part [that] rests in the cabinet of the late M. Hägerflycht." Liljewalch's work showed that this bone had a history of its own, which closely parallels the history of the skull in Paris. The problem is that this skull portion—the left parietal bone, which forms the left side of the crown of the head—is not missing from the skull in Paris. Both cannot be relics of René Descartes.

In 1983, the curved parietal bone in the display case in Lund caught the attention of C. G. Ahlström, a professor of pathology at the university, and together with two colleagues, he conducted a detailed scientific and historical investigation of it. Besides reporting anatomical particularities—measurements, coloring, a slight indentation on the frontal part of the sutura sagittalis (the joint between the two parietal bones)—they noted the fact that the bone is whole, which is suggestive. The human skull is not a single mass

but comprised of twenty-three separate bones, which are connected
by the ridged joints called sutures. The Lund bone is complete,
including its intricate sutures, which means it was not sawed off
or broken. "The totally intact character of the bone indicates that
it has been extracted from the cranium very carefully, probably by
using the so-called blast method," Ahlström and his colleagues
reported. "With this method the cranium cavity is filled with
dried peas or millet and is then filled with water and left to swell,
whereby the various cranial bones slowly separate from each other
through the increased intercranial pressure. The method, which
is still in use, has been applied for a very long time and was used
when preparing skulls from both humans and animals for natural
cabinets during the seventeenth and eighteenth centuries."

The curious picture that emerges, then, is of someone pains-
takingly pulling a skull apart in order to spread pieces around—to
multiply relics. Presumably what Arckenholtz got was another
piece of the same skull, which is now lost. Equally curious is the
fact that, beginning with Hägerflycht in the mid-1700s, several of
the successive owners of the complete skull *also* owned the skull
piece. The records of Lund University indicate that the parietal
bone entered its collection in 1780, as a gift from a woman whose
maiden name was Stiernman. Stiernman was also the name of
one of the owners of the complete skull, and it turns out that the
donor was the wife of Bishop Olof Celsius and daughter of Anders Anton Stiernman. This together with the information from
Arckenholtz—that in 1754 Hägerflycht had owned the skull
piece—means that three successive owners of the complete skull
(Hägerflycht, Stiernman, Celsius) also owned the separate parietal bone. There's no telling what they might have had in mind.
Perhaps the first, Hägerflycht, obtained one, which he took to be
genuine, then happened upon the other, which he bought out of

amusement, since one or the other piece was obviously not what it was purported to be. Or, since each by this time came with a pedigree, maybe he was hedging his bets in owning both, embellishing his cabinet of curiosities with a double curiosity. His collection, then, with its Descartes anomaly, would have passed in toto to Stiernman, and from him to Celsius. Then Celsius's wife—who maybe was creeped out by the whole business—gave the parietal bone to the university, while the skull went to the paramour Fischerström and so onward.

In the original 1780 entry recording the parietal bone in the Lund collection its authenticity was taken for granted, but very quickly doubts crept in. If it had spent years beneath the ground, why was it so pearly white and unblemished? How could the skull of a man nearly fifty-four years of age be so papery thin? Keepers of the Lund collection gradually distanced themselves from claims that it had actually been part of Descartes' skull. Hampus Cinthio, who gave me a tour of the collection, only chuckled at the notion.

A succession of Swedes ranging across the first century and a half of the modern era had, apparently, been taken in. Was it Planström's doing? Or was he guilty only of swiping the true head and holding on to it as a personal charm? The parietal bone doesn't appear in the historical record until a century after Descartes' death. One hundred years is too big a curtain even to guess what may have gone on behind its heavy drape.

IN 1821, THE ACADEMY OF SCIENCES in Paris—which was by now used to voting up or down on the legitimacy of whole branches of science—had taken considerable time over a single skull, and there seemed to be a consensus. The chemist Claude-Louis Berthollet wrote to his friend Berzelius in Stockholm after the first meeting over the skull, telling him that "the Academy of Sciences received last Monday with a religious reverence and a vivid sensibility the present that you sent. We compared the skull with the portrait of Descartes and recognized a correspondence between them that, together with the proof that you have brought together, leaves no doubt of the personage to whom this head belonged."

Berzelius wrote back thanking him for the information and then launching into a lengthy discussion of his work with alkaline sulfur and whether "in alkaline sulfur liquids the sulfuric acid forms before or after the addition of water." In his reply, Berthollet noted that at a follow-up meeting of the academy, Delambre had done his best to discredit the skull as being that of Descartes. But "his observations seemed not well founded," Berthollet observed.

The world's greatest assembly of scientists had reached a conclusion, one that rested not on an ideal of certainty but on the modern notion of probability. They had applied their doubts to the very head that had introduced doubt as a tool for advancing knowledge. And in the end they gave the head a nod.

Cranial Capacity

SOMEWHERE AROUND 1767, A BOY IN THE VILLAGE of Tiefenbronn in the German state of Baden observed that, among his classmates, those who had the best verbal memories—those who were most able to learn and recall lengthy passages of the Bible, for instance—tended to have bulging eyes. Franz Joseph Gall was nothing if not consistent throughout his life, and twenty years later, now a physician in Vienna, he began holding public lectures on what he called "organology." He had developed a new technique for dissecting the brain—not by slicing into it like a ham, as scientists had been doing previously, but by teasing apart and analyzing its separate structures—and based on this work he formed the idea that different parts of the brain controlled different types of mental activity. Had he stopped at that, Gall would have merited a noble spot in the canon of scientific pioneers, for the field of neuroscience is built upon the localization of brain functions, but he went further. He posited that the regions of the brain worked like the muscles of the body, so that those that were more highly evolved were more developed physically. That is to say, they bulged. By knowing the various regions of the brain and what mental faculties were located there, he reasoned, one could "read" someone's skull and determine his or her natural propensities.

It was one of Gall's acolytes who coined the term *phrenology.* Gall himself didn't like it, nor did he agree with many of the tenets of this new science of the mind as it was developed by others, but his is the name most associated with it. One reason is the aggressiveness with which he promoted it. His public talks and demonstrations in Vienna became wildly popular, as people grabbed on to this new, "modern" way to understand the human being in general and themselves in particular. Gall promised specificity: there were no fewer than twenty-seven brain regions in which individual functions or propensities—guile, courage, tendency to commit murder, sense of proportion, architectural talent, sense of satire, benevolence, obstinance, language ability (which indeed Gall situated behind the eyes)—were localized. One hundred and sixty years earlier, Descartes had all but proclaimed that science would unlock the mysteries of the human being; Gall was declaring that he had done it.

And, just as had happened with Descartes, Gall found himself opposed by the highest authorities in the land for promoting a philosophy that would undercut the established order. "This doctrine concerning the head, which is talked about with enthusiasm," wrote Francis, the emperor of Austria, in an edict forbidding Gall to continue expounding organology, "will perhaps cause a few to lose their heads and it leads to materialism, therefore is opposed to the first principles of morals and religion." The reasoning was much the same as what Regius encountered when he gave the first public lectures on Cartesianism in Utrecht in the 1630s. "Materialism" described a philosophy in which everything that goes to make up a human being is accounted for by material forces, thus leaving no room for theology. If goodness and a tendency to evil were somehow preprogrammed into the brain, what role was there for the church in governing human behavior? And since church and state were bound so closely in most of Europe (Francis had sent forces to oppose the French revolutionary government), such a categorical threat to religion was equally a threat to political power.

Like all good self-promoters, Gall made use of the controversy: he and his colleague Johann Spurzheim left Vienna and took their cranial road show on a thirty-city tour, becoming famous throughout Europe. When Gall arrived in Paris in 1807, crowds greeted him; the "head science" was caricatured in the press, and at parties young people began a semicomic fad of fingering one another's heads. Gall wanted to have it both ways: he relished the popular attention but he craved legitimacy, and the Academy of Sciences was indubitably the bestower of scientific legitimacy. He put an overview of his work before the academy in 1808. Its initial response was careful and mixed: in a tightly written fifteen-page report, the committee found Gall's anatomical work impressive but remained judiciously reticent on the subject of reading bumps.

Gall settled in Paris. He was determined to win official approval, and he continued to develop his ideas. The flaw in his organology theory was that it had little connection to his brain dissections. Then again, he relied heavily on a principle that ought to have endeared him to Georges Cuvier, who, as permanent secretary of the academy, was in a position to grant legitimacy to Gall's work. Cuvier was one of the founders of comparative anatomy, and this was the basis of Gall's argument. In Vienna, Gall had glimpsed the outlines of his localization idea while working at a mental hospital with patients suffering from one or another form of monomania; he reasoned that obsessive fixation on a certain subject or type of behavior might relate to a particular area of the brain. Later, working at a prison, he studied the heads of inmates and decided that he had discovered a common skull abnormality in most. This spot (just above the ear), he concluded, reflected a tendency toward criminality or antisocial behavior.

Comparing anatomical features would be the foundation on which Gall's theory would rest. In Vienna he made arrangements to have the police and the mental asylum provide him with the skulls of deceased murderers and "lunatics" so that he could analyze and compare them. Just as important, he believed, was to examine the skulls of people of worthy and notable achievement. Obtaining the heads of great thinkers, artists, and statesmen was not a simple matter, but Gall was persistent and over time he built a collection of three hundred skulls and plaster casts of skulls. The case of Goethe in particular highlights Gall's tactics and zeal in pursuing choice skulls, for Goethe was still alive at the time that Gall pursued him. Gall was so eager to have the skull of this particular genius that, not content with Goethe's graciously agreeing to have his head cast, he wrote to the sculptor who made the bust imploring him, in the event of the poet's death, "to bribe the relatives" to get them to give over the head for his collection.

Gall was obliged to leave most of his skulls behind in Vienna, but in Paris he promptly began a new collection, avidly pursuing notable examples. Events thus coalesced in a curious way in 1821. The same year that Gall decided he had amassed enough evidence and an impressive enough body of work to make his play for admission to the hallowed ranks of the Academy of Sciences, the skull of one of the most renowned of great thinkers—an object Gall would very much have liked to get his hands on—arrived at the academy.

Gall's chances of achieving the first objective were not good. The members were nearly unanimous in looking down on organology and "cranioscopy" (the palpating of the skull to assess a person's aptitudes and deficiencies). Cuvier in particular found that Gall's philosophy was speculative rather than based on clinical work; more importantly, as with evolution, he felt organology, with its assigning of so much of human intellect and emotion and predilection to biology, was an affront to what was to him the core, unimpeachable notion of an intelligent creator who bequeathed free will to his creations.

And indeed, when the question of his membership came up, Gall received only one vote, that of his friend the naturalist Etienne Geoffroy Saint-Hilaire. Bitterness did not, however, prevent him from continuing to beseech the academy. He had submitted the work on the basis of which his candidacy was to be judged on October 15, 1821, just one week after Delambre issued his final report on the skull of Descartes. Gall had probably been aware of the arrival of the skull five months earlier, in the package from Berzelius; he asked to be allowed to make a plaster cast of it. Cuvier granted the request, and Descartes thus became part of the Gall collection, alongside Voltaire and Goethe.

Gall died seven years later (of a cerebral hemorrhage, no less), having stipulated in his will that his own skull be added to the collection. It may seem curious, considering Cuvier's decisive rejection of the work with which the collection was associated, that after Gall's death Cuvier purchased the collection for the Museum of Natural History, but at the time even those who were opposed to the particular arguments of phrenology believed the comparative study of brains and skulls was worthwhile and could advance knowledge of the brain. It was in this same museum that the members of the academy had decided to place the skull of

Descartes once they had satisfied themselves as to its authenticity, so that the plaster cast of his skull now joined his actual skull as well as that of Gall, together with the museum's collection of bones of assorted primates and early homonids.

In 1821, while Cuvier was serving on the committee that reviewed Gall's membership, he had also been on a committee examining the scientific submissions of a rising young star in the field of brain research named Jean-Pierre Flourens. Flourens had started by following Gall but shifted decisively away, so much so that his career would involve a sustained attack on Gall and phrenology. Thus as Gall himself faded from the scene, his very head being tossed onto the data pile in the increasingly combative field of brain studies, a successor arose. Flourens was convinced that Gall had erred in a basic way by not grounding his theory in experimentation. Far from denying this charge, Gall had actually believed that the experimental method, because of its invasiveness, led to false conclusions. The hallmark of Gall's approach was observation. He sat and studied the labyrinthine structures of the brain, and he compared skulls. Flourens, by contrast, believed that scientists needed to be active in their efforts to unlock the secrets of the brain. He performed huge numbers of experiments on the brains of a variety of living animals (ducks, pigeons, frogs, cats, dogs), systematically identified their parts (cerebrum, cerebellum, medulla oblongata, and so on), then either removed the part whose purpose he wanted to understand or probed into the brain to stimulate the specific region (along the way, he also pioneered the use of chloroform as anesthesia), and finally studied the change in behavior of the animal. Antivivisectionism aside, such an approach would seem today to be a logical way for scientific research to be carried out, but Flourens had to defend the experimental method, and in particular he contrasted his approach with that of Gall. Observation alone, he said, was "too limited to be truthful." He acknowledged that experimentation could lead to false conclusions, but that only meant the experimenter had to be very sure of his method and be willing to revise his conclusions based on follow-up experimentation.

What is notable about Flourens's work—besides the pioneering study he made of the functions of the basic parts of the brain and his contribution to recognizing the importance of the experimental method itself—is the philosophical foundation from which he operated. For Flourens was something of a throwback among nineteenth-century scientists—a complete, unrepentant Cartesian. His many books are replete with references to Descartes, and he wrote, "I frequently quote Descartes; I even go further; for I dedicate my work to his memory. I am writing in opposition to a bad philosophy, while I am endeavoring to recall a sound one."

The bad philosophy Flourens had in mind was Gall's, and he meant not merely Gall's preoccupation with cranial bumps but what organology implied. If all the categories of human behavior and thought and aptitude corresponded, as Gall claimed, with particular corners of the brain, then the brain must be the mind. This may seem a not terribly notable distinction, but it turns out to be one of the thornier questions of modernity, right down to our day. As the Austrian emperor had feared, such a philosophy would seem to imply that the whole sphere of human action is reducible to the physical level, to bits of matter inside the skull. Humans, then, must be something like elaborate machines, whose functioning can in principle be completely understood and mapped out. Having taken up this view, one can relatively easily dismiss—or devalue or otherwise recategorize—not only the soul but much of human culture and civilization: art, religion, love, marriage and family ties, political and social relationships.

Ascribing thought and behavior to the level of biology seemed to take away the foundation from these things, making them mere ad hoc tools for dealing with life, which could be changed or discarded based on other criteria, such as what was most convenient for the individual. The possibility of undercutting social foundations caused alarm, so that there was—in the nineteenth century, as there had been in the seventeenth century with the first airing of modern philosophies—a quake or tremor in society.

Descartes, perhaps more than any other individual, had set this in motion, in part by choosing to analyze the human body as an object, just like anything else in nature. But Descartes staunchly fought back against critics who accused him of "atheism" (a catch-all term for materialism and all it might imply). Because his philosophy was built on a hard distinction between mind and body, and because he included soul in the concept of mind, he believed that rather than draining the meaning from humanity he had in fact maintained the separate integrity of the mind-soul while allowing science to work on the physical side of things.

Flourens followed Descartes in arguing against identifying the brain with the mind. His reasoning seems a bit curious, for brain-equals-mind would seem to be the logical conclusion of someone doing an anatomical study of the brain. You prick this nerve and it causes a contraction in that muscle; you tweak another area and see that it affects speech, or color, or awareness of right and wrong; eventually, seemingly inevitably, you come to believe that you have mapped out, within the physical masses encased in the skull, all of the attributes we associate with the mind. But like Descartes, Flourens believed, despite all his investigations of the brain, that mind was somehow *other*. Flourens argued that Gall's organology, with its localizing of functions in various parts of the brain, was little more than a gimmick to please the crowds that came to hear him lecture and watch him run his fingers over people's heads. Gall's separate functions, or faculties, weren't really physically distinct: "Your *faculty*," he taunted rhetorically, "is only a *word*."

Flourens recognized that Gall had garnered a great amount of popular momentum for his theory, and he believed that phrenology was bad science and that he, on behalf of the academy, had to stop it. "Each succeeding age has a philosophy of its own," he said. "The seventeenth century enthroned the philosophy of Descartes; the eighteenth that of Locke and Condillac; should the nineteenth enthrone that of Gall?" Elsewhere he tightened the comparison in a way that suggested a double tragedy: "Descartes goes off to die in Sweden, and Gall comes to reign in France."

Flourens insisted that the mind was not a collection of faculties but a single, whole, indivisible entity. In this, too, he was following his hero. A chief difference between mind and body, Descartes had written, is that "the body is, by its nature, always divisible, and the mind wholly indivisible." This conclusion was based on Descartes' observations of himself: "When I contemplate my own self, and consider myself as a thing that thinks, I cannot discover in myself any parts, but I clearly know and conceive that I am a thing absolutely one and complete."

This notion seems decidedly prepsychological. One feature of the modern world is precisely that we think of the self or psyche or person as composed of different parts, the names of which vary with the generations—ego and id, inner child, flower child, father figure, earth mother, Oedipus complex—so there's a sense in which, for all the absurdity of phrenology, Gall was the more modern thinker. His organology *was* a psychological system, a scientific attempt to analyze the individual that predated Freud by nearly a century.

But Flourens, along with the rest of the scientific establishment, was not ready to march in the direction that Gall indicated, for reasons that were both scientific and unscientific. The historian of science Robert M. Young underscores the divide between Flourens' science and his personal philosophy, writing that "Flourens' advocacy of physiological experimentation is complemented by a complete unwillingness to apply the scientific method to the study of mental phenomena." For all his commitment to the methods of science in the physical workings of the brain, Young goes on to say, "Flourens was not prepared to submit the human character, the mind, or its organ to analysis. Their unity was a necessary basis of his beliefs about man's dignity and freedom." Flourens was eager to slice the brain apart, but not the mind, and he was unwilling to do the latter because he believed it would lead to a breakdown of civilization.

But if there are political, social, and spiritual crises that come with equating mind and brain, there is an even more elemental problem with keeping them separated. Descartes, too, insisted on a complete split between the two, saying that the physical and the mental are distinct substances. The question that was thrown at him almost at once was that if body and mind exist in different universes, so to speak, how do they interact? How does your stomach's hunger transmit itself to your mind, and how then does your mind tell your legs to walk to the refrigerator and instruct your hand to open the door and direct your eyes to scan the shelves and cause your fingers to reach toward the slice of leftover pizza? In short, how does anyone ever do anything? If so basic a question brings this "mind-body dualism" theory to a halt, then there must be a serious flaw with the theory.

Descartes' solution, based on his own dissecting work, was to identify a small, nut-shaped structure in the center of the brain called the pineal gland as the place where the two came together:

"the principal seat of the soul," he called it, "and the place in which all our thoughts are formed." His reasoning had a rather disarming simplicity to it, which was based on symmetry. "The reason I believe this," he wrote,

> is that I cannot find any part of the brain, except this, which is not double. Since we see only one thing with two eyes, and hear only one voice with two ears, and in short have never more than one thought at a time, it must necessarily be the case that the impressions which enter by the two eyes or by the two ears, and so on, unite with each other in some part of the body before being considered by the soul. Now it is impossible to find any such place in the whole head except this gland; moreover it is situated in the most suitable possible place for this purpose, in the middle of all the concavities; and it is supported and surrounded by the little branches of the carotid arteries, which bring the spirits into the brain.

Descartes had scarcely aired this notion before critics pounced. If "mind" and "body" were truly distinct, how could a physical gland be a conduit of mental energy? This has been the critique ever since, and it does indeed reveal the absurdity of Descartes' effort to join mind and body, but it is worth noting that Descartes did not state categorically that he had solved the conundrum. In fact, not long before sailing off to Sweden, he admitted it might be too vast a problem for the mind to grasp: "It does not seem to me that the human mind is capable of forming a very distinct conception of both the distinction between the soul and the body and their union; for to do this it is necessary to conceive them as a single thing and at the same time to conceive them as two things; and this is absurd."

This rather uncharacteristic burst of modesty aside, Descartes gave dualism its modern form and insisted on it, and Western philosophy and the Western tradition since his time—modernity, in other words—have had the mind-body problem in their DNA, as it were. The problem is so elemental and yet so sweeping that attempts to solve it today run across many disciplines, from computer science to neuroscience to psychology. As both the Austrian emperor Francis and Voetius, the theologian in Utrecht who opposed Descartes, feared would happen, much of the Western world has solved the problem of dualism by coming down on the material side of the equation. Physicalism is the present-day term for the view that the physical or material world is the real world—that nothing exists outside of it—and a lot of scientists and philosophers ascribe to some form of it. People who declare themselves to be atheists and say that what they believe in is science or the physical world or the here and now are adopting a physicalist stance.

Jean-Pierre Flourens tried to hold the line against physicalism by bringing Cartesianism—which had long since receded—into cutting-edge nineteenth-century science. While his effort may seem disingenuous, as though he were putting on blinders to keep from looking at things that were too troubling to his worldview, there was a certain wisdom in his attempt. For, as many current thinkers have pointed out, there are basic problems with the physicalist view. What it leaves out, to state it briefly and bluntly, is me. The present-day philosopher Thomas Nagel puts it this way:

> For many philosophers the exemplary case of reality is the world described by physics, the science in which we have achieved our greatest detachment from a specifically human perspective on the world. But for precisely that reason physics is bound to

leave undescribed the irreducibly subjective character of conscious mental processes, whatever may be their intimate relation to the physical operation of the brain. The subjectivity of consciousness is an irreducible feature of reality—without which we couldn't do physics or anything else—and it must occupy as fundamental a place in any credible world view as matter, energy, space, time, and numbers.

That is to say, human consciousness is the well from which we derive much that is most meaningful to us, so any theory of knowledge that does not take it seriously into account—along with all of the stuff that goes with human consciousness: mourning the dead, petting kittens, bowing to Mecca, cherishing faded love letters, risking your life to save someone else's, subconsciously loathing your mother or consciously hating your boss—is flawed. This is the problem that people today have who decide to solve the puzzle of modernity by rejecting past systems—usually religious systems—and replacing them with a good, firm, "scientific" way of understanding. The classic scientific perspective is one of objectivity, and as Nagel says, "Although there is a connection between objectivity and reality . . . still not all reality is better understood the more objectively it is viewed." We ourselves—our individual consciousnesses, the very minds that seek an objective view and, having found it, try to hold on to it even as they are bombarded with thoughts and pains and desires—have to be made part of the picture.

The hard fact of modernity is that from the time that Descartes separated the two, nobody has yet come up with a definitive, universally satisfying way to solder mind and brain together again. Descartes declared in 1646 that it may not be possible. In 1998, Thomas Nagel stated flatly that "no one has a plausible answer to the mind-body problem." In 1808, when Cuvier led the committee that reviewed Gall's first effort to win the approval of the Academy of Sciences, he said much the same thing in his report. The brain, Cuvier and his colleagues wrote with considerable elegance and sophistication in their critique of Gall's science, seems somehow fundamentally different from the rest of the body, so that

> we cannot expect a physiological explanation of the action of the brain in animal life comparable to that of the other organs. In these other organs the causes and effects are of the same kind; when the heart causes the blood to circulate, it is one motion that produces another motion. . . . The functions of the brain are of a totally different order; they consist in receiving, by means of the nerves, and in transmitting immediately to the mind the impressions of the senses, in preserving the traces of these impressions, and in reproducing them . . . when the mind requires them, [and] lastly, in transmitting to the muscles, always by means of the nerves, the desires of the will. Yet these three functions suppose the always incomprehensible mutual influence of divisible matter and the indivisible self, an unbridgeable gap in the system of our ideas and an eternal stumbling block to all our philosophies.

There was then, as there is now, what might be termed a liberal-conservative divide in attempts to resolve the problem. Put another way, there is a connection between the esoteric efforts to tackle dualism and the sorts of real-world battles that fill newspapers and occupy TV talk shows. Those on the left have tended to accept the seeming consequences of equating mind and brain: if it means that basic features of society—the self, religion, marriage, moral systems—need to be reconstructed along new lines, so be it. Examples of such renovations in human values might include promoting equal rights for women and minorities, legalizing abortion, advocacy of same-sex marriage and gay adoption, and viewing other cultures and religions as being as valid as one's own. The point is not that mind-equals-brain requires one to hold particular positions on these topics but that it allows for a wide range of moral speculations. The "conservative" stance has been to fight to keep "mind" separate from "body"—to preserve the status quo, whether in matters of religion, the family, or the self, to maintain that there is an eternal, unchanging basis of values. With regard to Descartes, the irony is that the man who was once seen as the herald of the modern program, the breaker of all icons and traditions, had by the nineteenth century become part of the conservative argument, the man who built a protective wall around the eternal verities, keeping them from the corrosive forces of modernity.

The nineteenth-century version of the culture wars was fought on several scientific fronts. First, thanks to the continuing refinement of microscopes, was the development of cell theory, the idea that all living creatures are composed of basic structural units, cells, which divide themselves to form new cells. For materialistically inclined scientists and philosophers, the discovery of these building blocks of life did away with the need for metaphysical crutches; life was the result of a series of intricate physical interactions. The second popular front that opened in the cultural melee of the nineteenth century—Darwinism, and in particular the idea that humans are descended from apes—became the most infamous. Equally as elemental and contentious as cell theory and Darwinism was the area of brain research. At the French Academy of Sciences, Georges Cuvier—who as the permanent secretary was a gatekeeper of science and the established order and who had fought against the theory of evolution in part on religious grounds—seems to have looked on the arrival on the scene of Jean-Pierre Flourens as a godsend. Flourens was a brilliant, incisive scientist, but also one with a deep commitment to preserving the integrity of the mind and thus of the established social system. Cuvier took him up as his protégé and almost at once began grooming him as his successor. In 1832, as Cuvier lay dying, he urged his colleagues to choose Flourens to follow him as permanent secretary of the academy, and they did.

Cuvier died on May 13 of that year, and three days later, also following his wishes, Flourens oversaw an event that would be remarkable if carried out today but was somewhat less so then. A team of Cuvier's colleagues from the academy conducted an autopsy of his body. It was fashionable in the early nineteenth century for scientists to nominate friends to do postmortem examinations of them in the hope that even after death they would continue to advance scientific knowledge—a trend that reached its climax in 1875 with the founding in Paris of a "Society of Mutual Autopsy." Cuvier's friends duly sliced open his chest and belly and studied the vital organs there. Then they moved to the feature attraction. As the newspaper *Journal des débats* reported on its front page on May 16, "At the opening of the skull, all the assistants were struck by the development of the cerebral mass and above all by the truly prodigious number of convolutions present on the surface of this enormous brain."

Indeed, the brain of Cuvier—which during his life had played such a prominent part in the development of science, not to mention in assessing the authenticity of the skull of Descartes—would soon become quite literally involved in the next debate on brain science. The man who more or less invented the field of comparative anatomy was about to have his own anatomy put up for comparison.

I N THE AUTUMN OF 1857, while Charles Darwin was in London working feverishly on the manuscript of *On the Origin of Species,* a man named Pierre-Paul Broca visited a farm in the western French city of Angoulême and touched off what would be a smaller scientific storm than Darwin's but one that prompted the same clash of worldviews. He had come to see a farmer called Roux—or rather to see his animals. Roux had been crossbreeding hares and rabbits for some time and selling them to local butchers. Getting the two species to mate had been tricky— male hares turned out to be sensitive creatures that engaged in lengthy foreplay and this confused the female rabbits, which were used to more brusquely straightforward treatment—but once the farmer had figured out a way to ease the romantic differences he had had good results and a meat that pleased local sensibilities.

Broca had heard about these creatures and came to see for himself. The farmer had produced six or seven generations; Broca was intrigued, and undertook a scientific study. The animals had some features of the rabbit, some of the hare, some entirely their own. They reproduced. Broca wrote papers in which he pressed an argument: this was clearly a new species, which meant that scientists had to admit that "the classical doctrine of the permanence of species is entirely mistaken." Broca presented his conclusion at the Biology Society in Paris, where it was met with awkward silence. Darwin's book—and the writings of Alfred Russel Wallace, which would appear at about the same time—would cause an upheaval for arguing that evolution, across millennia, had produced the myriad varieties of life on earth. Here Broca was not only calling for a similarly revolutionary change in beliefs, but insisting that a new species had come into being in a matter of months. Of course, farmers had been practicing animal husbandry, selectively breeding their livestock, for centuries, but Broca was stating in declarative scientific language that the classic doctrine—Cuvier's "fixity of the species"—was nonsense. It also didn't help that he was speaking frankly about sex and sexual behavior.

The savants who headed the Biology Society applied pressure for him to desist, whereupon Broca decided to found a society of his own. Together with several other intellectuals, he had been working toward a new approach to science, or rather a new application of it, which would bring various disciplines together to study humans and human society. The idea was to treat the human being as dispassionately as any other animal species and try to understand it individually, in groups, and in its environment. Broca himself had studied medicine, but his preternatural energy and curiosity drove him in a variety of directions almost simultaneously, so that he became a surgeon, anatomist, brain researcher, cancer investigator, proponent of evolutionary theory, student of fossil records and spinal cord injuries, pioneer of blood transfusion techniques, and theoretician on the mechanisms of speech. His initiative following the rabbit-hare episode was to launch a new association—and with it a new field—that would bring together as many of these diverse arenas as possible. *Anthropology* was the term that had come into being for this multifaceted approach to humankind.

Broca's Anthropology Society ran into immediate problems. Broca needed government approval to found a scientific society, and to those in power this proposed new field, which would treat people as animals, seemed shockingly demeaning. Through indefatigability and by working the angles in various bureaucratic departments, Broca eventually got wary approval for his Anthropology Society to meet, on condition that a policeman sit in on the sessions to make sure they were not subversive. (For two years a policeman slumbered through the arcane discussions, then quietly stopped coming to the meetings.) Having gotten the sanction, he then proceeded to oversee gatherings in which a sweepingly diverse range of subjects were focused—largely by his own will—into coherence. The business with the rabbits and the hares was hardly tangential. Broca was serious about looking at human interactions fully and dispassionately, down to the level of mating and reproduction. This approach continued to scandalize some of his fellow scientists, but unlike, say, Cuvier or Flourens, Broca was politically and temperamentally a defier of the status quo, a hater of absolutes and establishments. For standing up against religion and superstition, as well as for trailblazing new fields of inquiry, he would become a hero to later scientists.

At the same time, he would become a trailblazer in quite a different way, one with few heroic associations. For in involving itself so intimately with human groups and their differences, the new field of anthropology became, almost at once, a kind of scientific institutionalization of the principles of racism. Certainly much of the work that the new society covered seems very similar to what an anthropology course today might focus on. A meeting in 1861 began with a lively study of "a *pilou-pilou*, or festival mask of the New Caledonians," whose carved shape incited great interest for being so unusual and yet resembling masks from ancient Greece. The same meeting featured a report by a member analyzing marriage to near relations in Ohio, where a law had recently been passed barring marriage by first cousins. The data, which the French anthropologists found to be "of high interest," showed that the 873 known marriages between first cousins in that state had produced 3,900 offspring, 2,400 of whom "were afflicted with grave deformities or complete imbecility." Custom and morality had long barred such unions; the Anthropology Society wished to explore the underlying genetic reasons for these prohibitions.

At the 1861 meeting, the society also received the skull of a New Caledonian tribesman, the skull of a native of the New Hebrides, a catalog comparing the skulls of different peoples, and a manuscript French translation of a book entitled *The Forms of the Head in the Various Races*. Given the focus of the nascent field, comparing cultural differences was a natural component of the work. But the comparisons weren't restricted to masks and they had a heavy weight attached to them. Discussion of origins and national and cultural identity reflected a desire to believe—to know—that one's people were somehow superior. Nineteenth-century science, in other words, was asked to confirm ancient self-image. In the case of France, the burning question was: Who were the original French people, and in what ways were they unique? Members of Broca's Anthropology Society held forth on the ancient Gauls and their relationship to Celtic peoples; they discussed Julius Caesar's *Conquest of Gaul* and teased out language clues in an effort to advance the argument of an originating race that gave France its distinctiveness. Along the way, the scientists developed any number of racial typographies featuring different breakdowns. One savant divided humankind into five racial categories: white, yellow, red, brown, and black. Another found fifteen flavors of humanity.

It's curious that, on the one hand, the field of anthropology was the clearest expression yet of a scientific commitment to objectivity, with its practitioners vowing to bring the dispassionate mode of inquiry that had become the hallmark of science deep inside the human being and the human community and, making fullest use of the Cartesian method, valiantly discarding traditions and ignoring the received wisdom of societies, nationalities, and social classes. On the other hand, the anthropologists became racists by definition as they threw themselves headlong into fraught questions regarding human types.

It would be a mistake, however, to think that these men stacked the deck against other races as they sought to delineate the superiority of white Europeans. The fact is they never even tried to set up a dummy "I wonder which race is superior?" question and then proceed to solve it. They weren't out to prove that whites were superior for the simple reason that they assumed it from the start. It was so obvious to them it didn't need proving. The proper use of science was in figuring out why this was so. Of course, racism was built into nearly every culture in the nineteenth century, and in partial defense of the field of anthropology in particular, it might be worth noting that in working through these beliefs— beginning to question the underlying superiority complex as an awareness of it slowly rose to the surface over the course of a century—the field helped to lead the way to the wholly new concept of racial equality.

Even more than on linguistic or historical clues, the nineteenth-century scientists relied on bodily characteristics to explore the question of how it was that their race was superior. Hair color, skin color, size and shape of teeth, nostril width, lip thickness, jawline, chest hair, breast size, penis length, sexual appetite—the level of detail on which they focused was graphic and encyclopedic.

Of all the parts of the body that came under scrutiny, the face and head were deemed foremost in importance. The so-called science of physiognomy, which was developed by an eighteenth-century Swiss poet named Johann Lavater and adapted by many eminent scientists of the nineteenth century, was built around the idea that features of the head and face of each human race correlate with intelligence and inclinations. The man who refined physiognomy and gave it scientific legitimacy was none other than Georges Cuvier. Some of Cuvier's investigations in comparative anatomy are impossible to read today without cringing. In *Lessons in Comparative Anatomy*, a textbook, he measured the facial angle—the angle from the forehead to the front of the teeth—in a variety of apes as well as in different racial categories of humans. A macaque had a facial angle of 45 degrees—a severe slope downward from the forehead to the lips. Monkeys were in the range of 60 degrees. An orangutan measured 67 degrees; this, for Cuvier, was meaningfully close to the angle of 70 degrees he found for the "Negro" and quite removed from the 80 degree angle in European skulls. The more extreme slope in the Negro and the orangutan Cuvier related to a flattened frontal bone and reduced size of the frontal lobe of the brain. A more severe facial angle, he went on to reason, meant a lesser intellect and a brain that was more driven by "animal" instincts. Add the facial angle to other comparative data and the full picture emerges: "The Negro race is confined to the south of the atlas," Cuvier wrote in a crisp overview. "Its complexion is black, its hair frizzy, its cranium compressed, and its nose crushed; its prominent muzzle and its large lips obviously bring it closer to the apes: the tribes that compose it have always remained barbarous."

Such data and conclusions became part of the foundation from which the science of anthropology began its investigation of the human animal. Another concept of comparative anatomy that featured prominently—and whose development came to a head, as it were, at the very moment that Broca founded his Anthropology Society—was cranial capacity. One component of Gall's work that had been accepted as relatively noncontroversial by the academy was the notion that the size of the brain (and thus the size of the skull) correlated with intelligence. That is to say, having a bigger brain meant that you were probably smarter.

While phrenology itself had been discredited by the French Academy of Sciences, it had continued to develop in the popular consciousness elsewhere. It had a lively run in England and the United States through much of the nineteenth century. In particular, it fit with American ideas of individualism and upward mobility. Gall's theory of localization of mental functions in the brain may have meant that some people were born with greater capacities for love or intelligence or wisdom than others, but everybody already knew that. What struck a nerve was that it seemed to show that social status didn't matter. Among parents, doctors, and educators in nineteenth-century America, phrenology thus evolved into a self-help program, whose tenets were that each person was endowed with certain strengths and weaknesses and that it was possible, through hard work, to improve.

At the same time, following Gall's formulation of his ideas in Vienna, phrenology had evolved in Germany along a parallel yet quite different social track. It became a force in the German wing of the culture war. Politically, Germany in the early nineteenth century was a confusing confederation of semifeudal states. In the 1840s, a movement to throw off the monarchical paraphernalia, embrace democracy, and unite into a federation caught fire across social classes. Phrenology actually became a plank in the political platform of a lawyer named Gustav von Struve, one of the leaders of the movement, and helped to foment a revolution in 1848. As in the United States, von Struve and others took from phrenology the notion that—the multitude of individual differences aside—

human brains, and thus minds, were essentially all the same. All, therefore, were entitled to the same status and treatment.

The revolution left a confusing political situation that from the point of view of traditionalists was only exacerbated by the growing acceptance of science and materialism. By midcentury German scientists were openly proclaiming atheism as a creed. Beginning in 1854, the situation led to a showdown between two renowned German scientists—physiologist Rudolf Wagner and zoologist Karl Vogt—who engaged in a public debate on whether the growing mountain of scientific data from all fronts contradicted biblical accounts of the origin of life. Wagner came down firmly on the conservative side: he believed that every scientific fact would be found to be in accord with the story of creation; he further argued that religion and religious morality were the foundations of society and that science had a duty to keep its investigations within the boundaries that Christian teaching allowed. Science, he warned his colleagues, "would be suspected of destroying the moral foundations of social order" if it developed a materialistic philosophy. Wagner identified the root of materialism in brain science, and he went so far as to claim that scientists in Germany had a patriotic duty "to the nation" to keep mind and brain apart. Vogt, a defiant materialist, ridiculed such ideas, declared it obvious that the brain was the organ of thought, and set out the equation of mind and brain in a particularly graphic analogy: "The brain secretes thoughts, as the stomach secretes gastric juice, the liver bile, and the kidneys urine." Later, Vogt took apart the Cartesian notion that equated nonmaterial soul and mind. "The activities of the soul are only functions of the brain," he asserted. "There is no independent soul."

Wagner found this sort of thinking dangerous nonsense yet, curiously enough, he reacted to it by devoting the last part of his career to a study of the doctrine of cranial capacity. His plan was to debunk materialism by taking on one of its most elemental notions. Surely if it was true that there was a firm link between the physical brain and seemingly nonmaterial thoughts, size would have to matter—that is, if the brain was the mind, then a bigger mind would necessitate a bigger brain. Therefore, he would do the most thorough study yet of the brains and skulls of eminent thinkers and compare them with those of normal people.

Wagner began his study in a fairly creepy way. As chance would have it, he had made his decision to examine great brains as Carl Friedrich Gauss, one of the greatest mathematicians of all time and one of Wagner's colleagues at the University of Göttingen, lay dying. The two men did not know each other well, but suddenly Wagner became a constant presence at Gauss's bedside. After his death, the family agreed to have his body examined for scientific purposes. Wagner got the brain.

Wagner went on to collect brains and skulls of other notables (a clutch of elderly thinkers at Göttingen died at around the same time, and Wagner managed to snag all of their heads), along with those of ordinary mortals, murderers, rapists, and the mentally ill, and wrote a study comparing weights and features. Wagner was forced to admit that the brains of geniuses seemed to have more convolutions on the surface, but he believed his tabulation of brains and their weights proved that size did not, in fact, matter. Of 964 brains, Gauss's, at 1,492 grams, came in in 125th place. Among those at the top of the list were brains that had belonged to an ordinary worker and "an imbecile."

The details of Wagner's study of brain weights reached Paris in February 1861 and instantly became the main topic of discussion for Pierre-Paul Broca's Anthropology Society. Broca himself had pioneered techniques for measuring skulls and brains and had become a believer in the correlation between brain size and intelligence. Far from being dissuaded by Wagner's paper, Broca found it close to proof of the theory. "Among the questions that have been put into discussion up to now in the Anthropology Society, none is equal in interest or importance to the current one," he told his colleagues. He went on to demonstrate that for him cranial capacity not only bore on the materialism debate but was tied to the effort to establish a hierarchy of races. "The great importance of craniology has struck anthropologists so much that many have neglected other parts of our science in order to devote ourselves exclusively to the study of skulls," he said. "This preference is without doubt legitimate, but it would not be if the examination of the bones of the head had only a purely anatomical significance and if one did not hope to find there some data relative to the intellectual value of the various human races."

Beginning in March 1861 and continuing for some months, Broca laid before his colleagues the evidence that both Wagner and he himself had amassed. To begin with, Broca acknowledged the unprecedented work that Wagner had done and the value that lay in it, but he found that Wagner's conclusions were entirely untrustworthy, owing to a methodological error. "Monsieur Wagner brought together pell-mell very disparate observations," Broca noted. "He enumerated, by weight, the brains of both sexes and all ages, the brains of idiots, epileptics, and the insane, hydrocephalics [those with 'water on the brain'], apoplectics, paralytics with or without madness; it surprises me that he confounded so many miscellaneous elements."

It was well known that women's brains tended to be smaller than men's and, of course, that children's were smaller than adults'; many of the ailments Broca cited had also been shown to result in either dramatically smaller or larger brains. Broca thus attempted to refocus Wagner's data by analyzing only what he believed to be the brains of healthy adult males. The new calculation brought the number down from 964 to 347. Also following Wagner, Broca supplemented the database by combing the historical record, some evidence of which he admitted was circumstantial. An account of the life of Pascal noted that after the death of the philosopher an autopsy was performed and the doctors observed that "there was a prodigious wealth of brain matter." An account from the autopsy of Oliver Cromwell reported that, according to Broca, "the brain of the Protector weighed no less than six and one-quarter pounds." Broca dismissed this figure as preposterously greater than that of any known brain, but he pointed out that in old English measurement there were only twelve ounces in a pound, so recalculating on that basis gave what was for Broca a truly massive yet just believable weight for the brain of the man who had launched the English civil war. Moving to another area of genius, Broca informed his colleagues that the brain of Lord Byron had weighed 1,807 grams, 400 grams more than that of the average adult male.

And what about French geniuses? Some very fine evidence, Broca asserted, lay virtually at their fingertips: the remains of the great Cuvier himself. "All the distinguished anatomists who assisted at the autopsy of Cuvier declared that they had never seen a brain so covered in convolutions so complicated and deep," he declared. Happily, those scientists had removed Cuvier's brain and weighed it, and Broca was pleased to report that the result—1,830 grams—put Cuvier at the very top of the list. In short, Broca's analysis of Wagner's tabulation—augmented and corrected by Broca himself—yielded compelling evidence of a connection between head size and intelligence. Broca summarized his conclusions succinctly: "In general, the brain is larger in men than in

women, in eminent men than in men of mediocre talent, in superior races than in inferior races. Other things being equal, there
is a remarkable relationship between the development of intelligence and the volume of the brain."

Cuvier himself would perhaps have had mixed feelings. On the
one hand, being used postmortem as part of a rigorous scientific
study would no doubt have pleased him. But he would not have
gone along with Broca's materialist thesis—much less with being
used in an effort to prove it.

Neither, as it happened, did one of Broca's esteemed colleagues
at the Anthropology Society care for Broca's argument. Louis-
Pierre Gratiolet was Broca's equal as a scientist and anatomist.
He was the first to recognize that the two hemispheres of the
brain each direct the movement of the opposite side of the body;
Gratiolet also identified the four lobes of the brain and gave them
their names (occipital, temporal, parietal, and frontal), as well as
the insula, a central region that is sometimes considered a fifth
lobe. Gratiolet happened to come from the same small town as
Pierre-Paul Broca (there is today a place Broca and a boulevard
Gratiolet in the town of Sainte-Foy-La-Grande), so that the two
men had known each other most of their lives. But Gratiolet had
had a hard time of it. He had been raised with expectations of
social position, but the family was always poor and despite education and achievements he had never managed the trick of getting
proper social standing, so that he stayed poor his whole life. Where
Broca was fiery and defiant, Gratiolet was quiet but indefatigable.
Broca commanded attention; Gratiolet, with his hunted, brooding aspect, tended to go unnoticed.

But Gratiolet knew the brain as well as anyone alive (as was
apparent in the detail, not to mention the title, of his magnum
opus, *The Cerebral Folds of Man and the Primates*), and as he pored

over Wagner's work he found himself unconvinced by Broca's recalibration of it, unable to make a connection between brain weight or skull size and intelligence. If Wagner had erred—perhaps intentionally fudging results by grouping disparate types of brains together in his table—it seemed to Gratiolet that Broca had also pushed and pulled the data to fit his theory. Since Broca put such emphasis on Cuvier's brain, Gratiolet delved into the Museum of Natural History in search of it, only to discover that the brain had not been preserved. Further, the skull was now missing.

Gratiolet came to the meeting of June 6 armed with this and other information. He took his colleagues carefully through Wagner's figures. Gauss was probably the greatest genius of the era, but the data, Gratiolet concluded, was clear: "This is not an enormous brain; this is not an exceptional weight." He took another example from Wagner's Göttingen brains: Johann Hausmann, a mineralogist. Seeing that his brain did not rank high on Wagner's list, Broca had demoted the man to a lower order of intelligence, but this didn't seem fair either to the data or to Hausmann. Germans who knew his work acknowledged that Hausmann was a scientist of great distinction, and French colleagues said so as well. "One can thus affirm without presumption that this was no vulgar mind," Gratiolet observed. "However, and I repeat: his brain is small."

Now it was time to get closer to home: Gratiolet challenged Broca on the size of Cuvier's brain. "If it is permitted to take [Broca's] expressions literally, the weight of Cuvier's brain is, if not monstrous, hardly less than supernatural," he began. However, there was no way to verify—or for that matter to disprove—the figure of 1,830 grams. But Gratiolet, in his dogged way, had pursued a novel idea. Weighing brains might not be a common practice, it occurred to him, but measuring heads was. And who measures heads but a hatter? He had consulted among former friends of Cuvier and, lo, he found one, a doctor by the name of Rousseau, who had in his possession a hat that had belonged to the former permanent secretary of the academy. Gratiolet took this item to one of the most renowned hatters in Paris, who told him that its size was at the large end of the scale but by no means excessive. But Gratiolet had more to add. "Cuvier," he told his colleagues, "had an extremely abundant mass of hair" that was furthermore quite bushy. Thus the dimensions of the hat were actually greater than the size of the skull it sat upon. "The measurements that I have just pointed out seem to prove rather obviously that if the skull of Cuvier had a considerable size, this size is not absolutely exceptional and unique."

But Gratiolet's presentation—his attempt to deconstruct the cranial capacity theory—had another chapter to it. Following the logic of the cranial capacity argument, a great intellect meant a large brain and skull, and a supreme intellect implied a supersized brain and skull. Gauss, a man of titanic mental achievement, had a brain of rather average size. Well and good: that was one example to counter the argument. Could the brain or skull of another genius of similar, or even greater, stature be found? Gratiolet knew the museum's collection well, and he produced just the item. If those favoring the cranial capacity theory had been discomfited by the example of Gauss, and somewhat mollified by that of Cuvier, what would they say to the head of René Descartes? Descartes had long since come to be regarded as not only the father of modernity but the intellectual father of the French nation. From his mind—from his *brain*, Broca and his cohorts would have insisted—had come the framework for approaching two centuries' worth of solutions or speculative solutions to particular riddles about the beating of the heart, the colors of the rainbow, and the setting of the sun—and indeed the nature of the mind. And yet, as had been immediately apparent to everyone who had handled it over the past 190 years, Descartes' cranium was a small and rather delicate object. The Latin poem written on its crown stated it plainly. *Parvula calvaria*: small skull.

"We possess in the Museum a skull that appears to be that of Descartes," Gratiolet began. "This skull, religiously preserved in Sweden in a circle of Cartesians, and covered with inscriptions that attest to its origins, eventually passed into more ignorant hands, so much so that one day it was sold for a vile sum in a public auction. Berzelius, who, by good fortune, assisted at the bidding, bought this precious relic and hastened to return it to France." Gratiolet had gotten a few details of the skull's wanderings wrong, but his main point was that the skull was in the museum's possession. It had been compared with a likeness of Descartes, he said, and the inscriptions on it had been checked. The academy itself had done the detective work, and had come away convinced as to its authenticity. "And yet, this skull, rather than being remarkably large, as it must be if genius depended on the volume of the brain, is, *au contraire*, quite small," Gratiolet said. But there was something else, which Gratiolet found compelling. This skull, he noted, "is admirably shaped." In fact, he went on, "This is one of the most beautiful types of Caucasian skull that it is possible to see, and, to summarize, I would say that it is the form and not the volume that gives the brain its dignity."

Gratiolet finished with the dramatic flourish of a prosecuting attorney who is convinced he has just demolished the opposition. There was an immediate response—from Ernest Aubertin, an anatomist and one of Broca's close associates—but it was more or less an "is too, is not" argument:

Monsieur Gratiolet . . . has said, if I have understood correctly, that the volume and the weight of the brain, whether considered among individuals or races, are of nearly no importance and that there is no connection between the development of intelligence and that of cerebral mass. This opinion appears to me erroneous. I don't pretend that intelligence depends solely on brain volume; but it depends in great part on it. As to the example of Descartes that our colleague has invoked one can oppose it with many others. Has it not been noticed many times that men of genius have enormous brains?

The "big brains" discussion unfolded over several months, completely enveloping the Anthropology Society, and Descartes' skull became a touchstone in the argument, to which members returned time and again. This was, as it were, the mind that had sketched the landscape of the modern mind. If brain equaled mind, then it simply was not possible that so great a mind could have fit into so small a brain receptacle. A month after Gratiolet brandished the skull of Descartes, Broca gave his thoughts on the subject:

Our colleague Monsieur Gratiolet knows the history of these famous brains; he thinks nevertheless that there is no link between the weight of the brain and the development of intelligence because, he tells us, there are some men of genius with mediocre brains. In support of this assertion he cited for us first the example of the skull of Descartes, then he invoked the research of Monsieur Wagner on the brains of some eminent men. . . . The skull of Descartes is without doubt a respectable relic, but it would be more still if it had a bit more authenticity. . . . Let us not forget, lastly, that the study of a skull, however complete, merely gives an approximate idea of the volume and above all of the *weight* of the brain. The brain of Descartes not having been examined, one will never know how to judge its worth. This example is thus without value.

A skull without a brain, Broca was arguing, is inconclusive, an empty shell. Yet as the discussion wound on, Broca and his lieutenants seemed beleaguered by the example of Descartes. On April 4, another member, a Monsieur Perier, mounted an attack on Gratiolet, that is to say on the skull of Descartes, that revealed that the mind-equals-brain camp had been doing some homework. "Monsieur Gratiolet has cited Descartes," Perier said,

> but without being able to affirm that the skull of mediocre dimension, which could have been his, belonged to the author of the *Discourse on the Method*. The relics of illustrious men are very much open to doubt. I know two cannon balls that are each said for certain to have killed Turenne [a French military hero] and that are preserved religiously, one at Sasbach, the other at the Invalides. And all those that travel in Switzerland know that it would be difficult to count the authentic crossbows that were used by William Tell to strike Gessler.

Perier then led his colleagues in a history lesson in which he recounted "the circumstances of the death of Descartes at the court of Stockholm," told how Christina had intended to bury him in the great hall of her ancestors, and observed that with such royal attention "one does not comprehend how the mortal remains of this famous philosopher could have been deprived of their most noble part." The head must therefore have been removed, he went on, "during the translation of the remains of Descartes to Paris and amid the hazards of this adventurous peregrination, which lasted more than eight months."

More to the point, Perier went on to compare the skull with what was considered the most authentic portrait of Descartes, by Frans Hals. Here, Perier said, one can plainly see "a large head, with a large and advanced forehead." Perier then read out the physical description of Descartes by Adrien Baillet, his seventeenth-century biographer, who talked of "the head a bit bigger in relation to the trunk" and a large forehead. These two historical records conformed with each other, and the skull matched neither. What was more, the historical images—in paint and in words—were in accord with what the learned men in the room believed to be the case, that "among minds or characters of great elevation" there is always extraordinary development of the brain and consequently an extraordinary size and weight. Perier then tossed Gratiolet's jibe at Cuvier back in his face by adding that "the hairstyles, as well as the helmets and the hats, of these privileged men, will be likewise exceptional."

The tireless Gratiolet came right back during the meeting of April 18, 1861, addressing his colleagues with a tone at once insistent and conciliatory. Some time earlier he had come to the conclusion that intelligence was not necessarily in proportion to the size of the brain. "However, when I expressed this opinion recently before the society it was not well received by my colleagues," he said. "It was vibrantly attacked, with rare skill and, I must add, with a courtesy for which I feel the need here to thank those who contradict me. Opposite such benevolent adversaries, to declare myself converted would hardly cost me. Unfortunately, I was not convinced. I still persist in my first thought, and, believing it still to be true, I will try to defend it and justify it."

He patiently went back through all of the main flaws he had found in the theory. Then he returned to the matter of Descartes' skull. Perier, he charged, had misread Baillet. Baillet had indeed described Descartes as having a head "a bit bigger in relation to the trunk," but he had also said that Descartes' body was "smaller than average." The net result could easily have been a smallish skull. As to the portrait by Hals, it was an excellent portrait, and while it showed a man with a rather large head in relation to his body, there was no way of relating it to other heads. "If this is not the head of Descartes," Gratiolet added of the skull, with a poke at his adversaries' intelligence-based argument, "then it could be that of his ignorant brother."

I T IS NOT ENTIRELY coincidental that, in the same week that Gratiolet countered the cranial capacity arguments of his colleagues in the Anthropology Society, off the coast of South Carolina American gun batteries opened fire on one of their own bases, Fort Sumter. Dating to the colonial period, slavery, race, and ideas of inherent racial differences had formed at the core of American society; during the same period, modernist ideas of racial equality were simmering. The tensions reached one climax with the attack on Fort Sumter, which signaled the beginning of the American Civil War. If racism was part of the American social fabric, it was also engrained in European intellectual life, and in the program of science as it developed. But modernity—going back to the seventeenth century and some of the "radical Enlightenment" figures who first made the connection between a commitment to reason and social equality—also seemed to have in it the possibility of working its way out of such dark corners.

For the record, mainstream science today has concluded that not only is there no appreciable correlation between race and intelligence, and no correlation between brain size and intelligence, but there is little genetic basis for the very idea of race. The dispersal of humanity and its division into various groups happened so recently, on an evolutionary time line, that racial differences are little more than skin deep. Variations beyond what the eye perceives, such as in intellectual capacity, involve a large percentage of the total number of genes and would require a far longer period of time to come about. J. Craig Venter, the geneticist whose company, Celera Genomics, became a private competitor to the Human Genome Project, put it this way to the *New York Times:* "Race is a social concept, not a scientific one."

Nevertheless, the impulses behind "cranial capacity," facial angle, and other literal forms of racial profiling are still with us. The best-selling 1994 book *The Bell Curve,* in the process of examining links between intelligence and economic standing, delved into possible links between intelligence and race, and its success suggested great pools of sympathy for such an argument. The psychologist J. Philippe Rushton promotes an updated form of the race-and-intelligence theory that he characterizes on his Web site in this way: "In new studies and reviews of the world literature, I consistently find that East Asians and their descendants average a larger brain size, greater intelligence, more sexual restraint, slower rates of maturation, and greater law abidingness and social organization than do Europeans and their descendants who average higher scores on these dimensions than do Africans and their descendants."

Rushton is not in the scientific mainstream, but James Watson—who won the Nobel Prize for codiscovering the structure of DNA—is. In 2007, Watson avowed a clear connection between race and intelligence, which he believes society refuses to acknowledge out of fear of political incorrectness. Watson said he was "inherently gloomy about the prospect of Africa" because "all our social policies are based on the fact that their intelligence is the same as ours, whereas all the testing says not really." He also wrote that "there is no firm reason to anticipate that the intellectual capacities of peoples geographically separated in their evolution should prove to have evolved identically. Our wanting to reserve equal powers of reason as some universal heritage of humanity will not be enough to make it so."

In response, other members of the scientific establishment chastised or ridiculed the seventy-nine-year-old Watson for being on this point, as one scientist put it, "out of his depth scientifically, quite apart from socially and politically." In *What Is Intelligence?* (2007), sociologist James Flynn analyzed intelligence tests and concluded that racial and ethnic differences in test scores—Africans tend to score lower than Western Europeans, for example—correlate with a society's familiarity with the principles of abstraction. Such tests, in other words, measure not necessarily pure intelligence but *modern* intelligence, the kind that expects us to critically analyze product labels and political speeches—and, for that matter, test scores. Steadily, over the course of a century, methodologies for linking race and intelligence have been tested and found wanting. That process might be said to have begun with Gratiolet's challenge to Broca. In a small way, Descartes' skull helped debunk bad science.

The immediate battle between Broca and Gratiolet ended with no certainty—with both sides believing they were right. But during the course of the debate—in fact, in the very meeting at which Gratiolet responded to Perier—Broca brought something new to the table. His attention had been diverted from the subject of big brains by a remarkable patient he had seen in his hospital rounds the week before, who had died shortly thereafter. The fifty-one-year-old man had been unable to speak for the previous twenty-one years, as it turned out as a result of contracting syphilis. Broca immediately devoted himself fully to this case, for speech was a basic part of human mental functioning and he felt sure that such a thorough loss must leave a mark on the brain. He performed the autopsy, removed the brain, and located a lesion on the left frontal lobe in a fold called the inferior frontal gyrus. This region, he quickly concluded, controlled speech.

Later work would show that he was basically right. The case study of "Tan" (the pseudonym used for the patient by Broca, after the only syllable the man had uttered) has become a touchstone in the history of science. The brain region Broca identified—now known as Broca's area—has been the focus of a great deal of research in neuroscience and speech disorders.

It may be a little easier to look back, on the nineteenth century, for example, to appreciate that useful and destructive approaches to problems—good and bad science—are always taking place simultaneously. Gall theorized the localization of brain functions while also promising to tell your fortune by reading bumps on your head. Cuvier laid the foundations of modern biology at the same time as he was engaging in lurid studies to prove black people were cousins of orangutans. Happily for Broca's reputation, his discovery of an area of the brain devoted to speech, rather than his work trying to relate brain size to intelligence, would become his claim to fame. Beyond that, the discovery of Broca's area was the first clear proof of the localization of brain functions. In one go, Gall was partially vindicated, science had advanced, and the materialists counted a score for their side. Of course, that score was in a game whose rules would continue to be challenged.

After Broca's death in 1880, his body was in turn subjected to a collegial autopsy. A thoughtful associate had the idea to etch his name—P.BROCA—onto the brain itself, on the inferior frontal gyrus—Broca's area. His brain then took up residence in the collection that he himself had amassed. That collection was eventually merged with the bone collection of the French Anthropology Museum, where Broca's brain joined the skull of Descartes.

6

Habeas Corpus

IF YOU HAD ASKED AN INHABITANT OF EARLY-twentieth-century Cleveland or London or Stuttgart to name the most modern city in the world, chances are the answer would have been Paris. Partly the belief was based on demeanor or sensibility: there was the café society, the commitment to discussion and debate, to art, to literature, to food and wine, and of course the open and knowing attitude toward sex. But demeanor was built on infrastructure; in Paris, more than almost anywhere else, steel and concrete and electrical currents had been put to use to provide a foundation on which modern life—longer, healthier, more comfortable and expansive and reflective than what previous generations could have hoped for—could base itself. The 1900 Olympics gave the wider world a window onto what a city could be. Visitors found a city in which the ancient past sat side by side with the future. The wide boulevards that Baron Haussmann had pressed through two generations earlier gave the city an open, contemporary feel. Paris had a subway, electric street lamps, elevators, a sewer system so modern and efficient people took boat tours in it.

A symbol of the city's modernity could be found at every major intersection: a wide-faced clock, high up on an ornate wrought-iron pole, its Roman numerals readable at night thanks to the lamp that topped it. Punctuality was to the late nineteenth and early twentieth centuries what standardized units of measurement were to the late eighteenth and early nineteenth and what computers would be to the late twentieth. It brought people into line with one another. Life regularized and became more orderly. These clocks were considered a modern marvel the world over because people knew that if the one at the place St.-Sulpice registered exactly twelve noon, the ones at the Trocadero and the place Vendôme and on the Ile de la Cité showed precisely the same time. The mechanism behind this magic was air. A central clock was connected to a machine that compressed hundreds of discrete charges; when the clock struck a new minute, the air bursts pumped out through miles of tubing each to their corresponding clock, where the thrust of pressure clicked off the next minute.

A little before eleven o'clock on the morning of January 21, 1910, the clocks of Paris stopped. Savage rains had fallen for weeks, and during the night the Seine had spilled over its banks, flooding streets and basements. The central clock room was inundated. Around the same time the clocks stopped, the street lamps went out and the subway cars came to a halt. The rain continued. In the poor areas in the east, neighborhoods became lakes, and the pressure from the water caused hundreds of buildings to collapse. Whole streets caved in from the weight. Outside the Gare St.-Lazare, water filling the underground train tunnels sent the street heaving upward, flinging pedestrians. All told, one million people had to flee their homes. Hospitals had to be evacuated. The Palais de Justice and the Prefecture of Police were flooded, hampering official efforts to deal with the catastrophe. Water poured into the basement of the Louvre, and curators scrambled to save artworks.

For more than two years the city struggled to right itself. State-of-the-art steam pumps were placed at strategic spots and drainage sewers were built to capture overflow waters in the event of future flooding, but in January 1912, with much of the city still largely stricken from the great flood of 1910, more torrential rains came, destroying the new sewers and submerging the city all over again. If, as Descartes had had it, the forces of modernity were dedicated to overcoming nature, with the floods of 1910–12 nature seemed to have struck back decisively; it was as though modernity itself had collapsed.

During the flood of 1910, the waters on the southern bank of the Seine had engulfed the quai d'Austerlitz, swept down the rue Buffon, and inundated the anthropological galleries of the Museum of Natural History, filling its rooms to a height of one and a half meters. The river had coursed through the collections of artifacts built up by the likes of Georges Cuvier, Pierre-Paul Broca, and Franz Joseph Gall and carried off skulls and femurs and rib cages on a grim tide. Like many other institutions, the museum took a long while to recover. By the fall of 1912, its galleries and storage spaces were still being renovated and many of the artifacts in its collections were stacked in piles.

Elsewhere, meanwhile, business was being conducted as usual, and September 23 was an ordinary day at the Academy of Sciences. The members received a report on a species of fresh water shrimp that lived in Lake Tanganyika, a work on paleontology, and a description of an astronomical device that would aid in determining the positions of stars. It might be expected that the most noteworthy item on the agenda would have been the reception of a publication analyzing the rainfall and flow of water in the previous two years. Instead, what got notice—not just in the chamber but in the press and all over the city—was an obscure volume that had recently been published in Sweden, a copy of which was on this day formally received by the academy: *The Correspondence between Berzelius and C.-L. Berthollet (1810–1822).* In poring through the pages of letters that the two chemists of a century earlier had exchanged, one of the members stopped short at the letter in which Berzelius noted that he had recently discovered the skull of René Descartes and sent it to the academy.

"Great turmoil, the 23rd of September last, at the Academy of Sciences!" began a report prepared by René Verneau, staff anthropologist at the Museum of Natural History, in which he tried to describe the ruckus that ensued after this arcane bit of information came out. With the passage of time, the fact that Cuvier had exhibited the skull for his fellow scientists had apparently faded from the institutional memory. Was it true, members wanted to know, that the skull of the great Descartes had been entrusted to the academy? If so, where was it?

After a flurry of orders and some days of searching in the various institutions that were affiliated with the academy, positive news came from the Museum of Natural History. Paperwork showed that the skull had been entrusted into the museum's keeping by Cuvier himself. The information was followed by an awkward admission: the museum was not at present able to locate the item in question.

At this, the matter turned into a news story. Reporters visited the museum and found piles of bones, unlabeled and unattended since the flood. Piecing together cause and effect, they wrote stories speculating that the skull of the intellectual father of the French nation had been swept away by the remorseless waters of the Seine. To make matters worse, the museum hadn't even known the item was missing. "This communication arouses definite emotion," said the *Journal des débats politiques et littéraires*. The *Gazette de France* waxed historical:

> One knew that the great philosopher died in Stockholm in 1650 and that his body was removed to Paris for inhumation. But it was not generally known that the body was missing its head. This had been, it seems, preserved, toward what end no one can say, by a Swedish officer. The skull of Descartes, duly labeled, was transmitted to the descendants of the officer and offered later to the Swedish Academy, which sent it to France. And does one find this relic today? It is believed to have been deposited in the museum, but there is no precise information in this regard.

Verneau felt the pressure that the news accounts generated. "Every hypothesis was allowed, and many publicists hastened to write up sensational articles in which were found the most fantastic assertions," he later complained. The museum staff conducted a thorough search of its collection. Verneau gathered the dossier on the skull, including Delambre's report and the rest of the documentation concerning the events of 1821, so that he and his staff were armed with information that would help them identify one skull from among hundreds. And at last an item seeming to correspond with the old reports—a skull missing only its lower jaw, covered with faded writing—was found, apparently in a jumble with other antique human remains, including some items from the Gall collection.

Within a week of the first news that there was such a thing as a skull of René Descartes in French institutional keeping, the presumed item itself was delivered to the desk of Edmond Perrier, the director of the museum. That same day, Perrier took it to the academy, so that its members, at their September 30 meeting, could see it for themselves. "This was the first time since 1821 that it left our national collections," Verneau said later, and, still smarting under the accusations of mishandling the object, he added, "and it was back in its place two hours later."

So ninety-one years after Cuvier's dramatic unveiling of the skull before the assembled scientific establishment of the academy, Perrier presented it again in the same room—"with all the respect due to this precious relic." He proceeded to give the scientists a brief history of the object. It seemed that it had in fact been on public display for some years, but "we stopped exposing it in our public galleries" because it came to be seen as improper to display the heads of people who had surviving family members. Perrier glossed over the temporary loss of the skull and its rediscovery in a heap of other bones, complaining instead about how since the flood there was no room to store anything properly.

But the second appearance of the skull at the Academy of Sciences did not end with quiet acceptance of the situation any more than did the first appearance. In fact, it sparked a whole new round of questions that engulfed the city as the skull became the topic du jour in cafés and parlors. "They talked about it . . . in Paris and in the provinces, on both poles and the two hemispheres; there was not another subject of conversation for twenty-four hours," a Dr. Cabanès wrote a month later in the *Gazette medicale de Paris*, which was hardly a sensationalist journal. "And this agitation, this tempest . . . about a skull? But it is no ordinary skull but the skull of one of our most illustrious philosophers, . . . the author of the *Discourse on the Method*, Descartes, no less!"

The issue was once again authenticity. The experts expressed doubts about the methods and conclusions of their predecessors of the previous century, and ordinary Parisians in turn took the matter seriously as well. If this was indeed the skull of René Descartes, it deserved to be kept as—in the words of the newspaper of record, *Le Temps*—"*la précieuse relique.*" But how did anyone know for certain, the editor of the highbrow journal *Æsculape* asked, whether "this vessel blackened by time has housed the highest thoughts or if it contained the lowly brain of some humble brewer"? Some sort of panel of experts needed to be called on, a collection of talents that ranged from historical to forensic.

The panel was found in the person of Paul Richer. He was a medical doctor who had worked under the renowned Jean-Martin Charcot, helping him develop his theory of hysteria; he was not only an anatomist and a member of the Academy of Medicine but also a sculptor and painter of great skill. Finally, Richer was an art historian with particular expertise in such matters as the anatomical correctness of Renaissance art—the muscles rippling beneath the skin in Michelangelo's sinewy *Last Judgment* figures, Raphael's soft, fleshy Madonnas. Remarkably, Richer was currently artist in residence at the Ecole des Beaux-Arts, which was housed in the former convent on the Seine where, more than a century earlier, Alexandre Lenoir had located his Museum of French Monuments.

Jean-Gaston Darboux, a mathematician who served as the academy's permanent secretary, asked Richer for help in addressing a unique problem. Richer spoke two languages fluently: art and science. He could be counted on to apply principles of both fields. Richer chose to begin by assuming that the skull was authentic. If so, should it not be expected to conform in its features with an authentic image of Descartes, such as a painting made from life by a master? It was well known that there was a portrait of Descartes by one of the great portrait painters of all time, the

Dutchman Frans Hals. This iconic image—of a brooding, rather dashing figure, a man of the world, a philosopher cum swashbuckler almost—hung in the Louvre, where indeed it still hangs. This was, Richer said, "the most accurate of the portraits of Descartes." Further, "a head presents a certain number of very precise points of reference in the bones that are particularly apparent in the painting by the Dutch master."

Richer began his analysis of the skull by having a technical draftsman who had not seen the skull in effect strip the flesh from the Hals portrait. That is, working from very sharp large-format photographs of the painting and employing a camera lucida—a device that uses mirrors to superimpose an image onto a canvas or paper—the draftsman created an extremely accurate drawing of the skull that he believed the man in the Hals portrait must have had, including every dent or protrusion, the height of the cheekbones, the breadth of the forehead, the set of the chin. Richer himself made a drawing of the skull of Descartes, being careful to execute it from the same position as that of the Louvre portrait—turning slightly to look to the right toward the artist—and on the same scale as the draftsman's drawing.

Before members of the academy and the press, in an atmosphere of unusual drama, Richer brought the two drawings together and superimposed them. Detail for detail, as he described it—"the receding forehead, the projection of the orbital arches some distance from the equally prominent superciliary arch . . . the width of the face . . . the projection of the nasal bones, which although broken at their end indicate an abrupt nose . . . finally the shortness of the naso-alveolar distance, which agrees with a certain brevity of the upper lip"—they were nearly identical. To sharpen the comparison, Richer had also had drawings made of other, random skulls in the same position. The audience saw that these did not correspond so neatly with the first two and in fact showed "notable discordances." Richer summed up with élan: "The skull preserved in the museum offers as absolute a similarity as is possible with that revealed in the portrait by Frans Hals."

Disparate events over the past few months had coalesced. The news of the apparent loss and then recovery of the exotic but iconic artifact had built over the course of the winter, with the city's recent troubles perhaps fueling a renewed concern for the French past. The atmosphere inside the academy for Richer's demonstration was so charged, and the demonstration itself so elegant, compelling, and definitive, that the result was a news story that traveled around the world. "Sketch Identifies Skull of Descartes" was the front-page story in the *New York Times*. "The Skull of Descartes Is Authentic," blared *Le Figaro*. The French newspaper went on to declare that "the method followed by the wise anatomist is a marvel from the point of view of scientific logic" and that "the result is conclusive: it is indeed the skull of Descartes that the museum possesses, and the documents and graphics put before the eyes of the academy leave no room for doubt." For Paul Richer, the article went on to say, the outcome represented "not only a personal success of his very scientific method" but the development of a method to be used in all future anthropological reconstructions.

The *New York Times*, after noting that "the results of Prof. Richer's investigations are generally regarded as settling once and for all a question of great historic interest," added that "a movement is now afoot for placing the skull together with Descartes's body, which reposes in the Church Saint Germain at Despres, in the Pantheon." *Le Figaro*, however, reported other plans: a case "would be constructed where the skull of the great thinker will be exposed, together with the documents that are proof, from now on unquestionable, of its authenticity."

The acclaim was universal. The case was solved. The matter was put to bed. Everyone on earth who had any interest whatever in René Descartes or philosophy or French history or skulls, it seemed, was satisfied.

With one exception. Some of the articles lauding Richer's success had picked up on an interesting point—that Richer's quest in some way mirrored that of Descartes himself. Richer—a sixty-three-year-old man with a bushy beard and lively, kindly eyes, whose studio at the Ecole des Beaux-Arts contained a selection of scientific instruments alongside clay models and art sketches—had all his life striven for accuracy, for an underlying method. What had motivated and united his medical work, his art, and his art historical study was a desire to get beneath the surface, especially the surface of the skin. Like Descartes himself, Richer wanted to understand the inner workings of the human body. And like Descartes, he was something of an aficionado of doubt. As the weeks wore on after his great success at the academy, what he came to feel was not that he had necessarily been wrong in his conclusion. Something else, equally Cartesian, nagged at him. Descartes' own method—*the* method—had begun not with unfocused doubt but with a questioning of received wisdom, of everything that was not clear and distinct to the individual mind of the thinker. The Cartesian notion that wormed into Richer's mind, even as the congratulations were coming in, was this: how did he know it was true that, as he had told the audience at the academy, "the most authentic of the portraits of Descartes is without contradiction that painted by Frans Hals, which is now owned by the Louvre"? As an art historian, Richer knew that attributions of works done centuries before float on very choppy seas. What evidence was there that the Hals portrait had been painted from life?

As it happened, Richer's doubts were in some respects justified. There is no reference to the other man in the respective biographies of Descartes and Hals; no document actually places the two men together. And the provenance of the painting in the Louvre is far from secure. When, in 1785, King Louis XVI bought a château west of Paris as a home for his wife, Marie Antoinette, the portrait of Descartes came with the package, it having previously been owned by the château's owners, the dukes of Orleans. But records further back are murky. The year 1649 had come to be associated with the portrait, so that it was generally held that Hals had painted Descartes then—the year before he died. But Richer came to believe that this was all mere supposition.

Richer's doubts were picked up by later art historians. In the 1960s, Seymour Slive, one of the great twentieth-century art experts and an authority on Frans Hals, took up the matter, observing that many portraits of Descartes, several of which look alike, seem to date from the same period. "Descartes was internationally mourned," Slive noted, "a fact that helps explain the number of portraits made of him soon after his death; judging from the number made during the following decades, the demand for them increased rather than diminished." Slive regarded the Louvre portrait—so smooth and serene and polished—as lacking Hals's lively brushwork and depth of character.

Could it be then that Hals and Descartes never met? Was the Louvre portrait by someone else? Had it perhaps been painted later, after Descartes' death? Was it even a portrait of René Descartes?

Paul Richer didn't have definitive answers, and for that matter few things in art history are certain, but evidence from another quarter sheds some different light. Baillet, Descartes' seventeenth-century biographer, noted that just before Descartes boarded a ship from Holland to Sweden in what would be the final voyage of his life, he was lured by a friend—a Dutch priest named Bloemaert—to pay a visit. Father Bloemaert lived in Haarlem, and while he had the great philosopher in town he begged him to sit for a portrait. Frans Hals spent nearly his entire life in Haarlem and was the preeminent artist in the city at the time Descartes visited. Of the many portraits of Descartes in existence, one struck Seymour Slive forcefully as having all the hallmarks of a Hals. It hangs today in the Statens Museum for Kunst in Copenhagen. The problem with it—if your interest is in having a definitive, iconic image of a renowned historical figure—is that it's very rough, not a finished painting at all, in fact, but a quick oil sketch; it's also a rather cloudy, murky image. Yet in the estimation of Slive and other, later experts, it has the vigorous brushwork characteristic of Hals, and a deeper sense of personality than the Louvre portrait. Even though he is posed, the man in this little painting seems to be in motion, as if he had been in the process of swiveling away until, caught by something, his eyes stopped to gaze into the viewer's. It is also a realistic, rather unkempt figure in this painting: there is some puffiness in the skin of the face; the man's hair looks dirty. He seems sad.

The Louvre portrait is still widely associated with Frans Hals—it is routinely identified in art textbooks as by Hals—but the Louvre itself now labels it as "after Hals," while the Copenhagen museum considers its Descartes to be by the Dutch master himself.

This reshuffling of attribution came after Paul Richer's time. So what relevance does it have for his investigation of the skull? If Slive and others are right, then Hals did indeed paint the philosopher—in Haarlem, at the time when Descartes, weary from his battles with theologians and others over what his philosophy would mean for religion, was about to sail off to meet Queen Christina, as well as his death, in the "land of bears between rocks and ice." And if they are right, the portrait painted from life is not the one in the Louvre—not the one Richer worked from—but the one in Copenhagen. Other portraits, including the one in the Louvre, seem to have been based either on this little oil sketch or on one of several earlier portraits that were painted of Descartes, none of which have completely solid provenance but any or all of which could have been done from life. As a check against the difficulty of knowing which of the portraits might have been painted from life, Richer repeated his process—having a draftsman create a drawing of the underlying skull—for several of these other portraits of Descartes. In all cases, in his estimation, the features on the portraits matched up with those of the skull—which to him suggested that, even if some of these paintings were copies, they were accurate in reflecting the bone structure of the face of the living man and that this structure matched that of the skull. So Richer had reconfirmation of his own work. He could put the matter to rest.

But as time went by the skull itself commanded Richer's attention, and he conceived yet another project. Immediately after Descartes' death, his friend Pierre Chanut had ordered a death mask made. This cast soon disappeared, but not before Christina commissioned her royal portrait painter, the Dutchman David Beck, to craft a postmortem portrait from it. The resulting painting is not the most lifelike, but Richer now hit on a use for it. This new project was a true obsession in the sense that no one else on earth paid much attention to it, for the obvious reason that, with world war in the offing, more important matters were afoot, but also because as far as others were concerned Richer had already established the authenticity of the skull of Descartes. Using his considerable skills as a sculptor, Richer fashioned a life-sized bust of René Descartes—noble of feature, classical in design—but a novel one, a sort of double bust. Starting with the Beck portrait, he had worked from the outside—the skin—inward toward the underlying bone. At the same time, he had created a plaster cast of the skull and worked from it outward, layering muscle and tissue until he reached the outer surface. What's more, the face on the finished sculpture was removable, and when it was taken away the cast of the skull was revealed, grinning its perpetual ghoulish grin. ·

The bust exists still in the collection of the Ecole des Beaux-Arts. As symbol or metaphor it has less to do with philosophy, art, or the human body than it does with one of the most persistent of modern preoccupations: the quest for certainty in an uncertain world. Or, to put it in the negative, Paul Richer's ghastly bust of Descartes is a manifestation of Doubt itself.

ONE OF THE ABIDING subproblems of the mind-body dichotomy, according to more than a few critics, is that it stacks the deck in favor of the mind. Think of yourself. Naturally your body—its aches and appetites—is part of the package, part of what you mean by "me," but in all likelihood it is the other layer—dreams, hopes, guilt, memories, invented memories, relationships, knowledge, mad stratagems, paranoias, e-mails sent, movies watched, battles to wage, remorse to chew—that forms the greater portion of your self-image. No doubt this is partly because the mind, as the entity doing the thinking, has a tendency to cast itself in the lead role.

Curiously enough, the same imbalance is in effect in the saga of Descartes' bones. The skull—that is to say, the material representation of "mind"—has occupied more people, has taken up more thought and energy, than have the rest of the remains. While the skull became the object of a detective story that encompassed some of the great scientific minds in history and has been subjected to artistic and scientific analysis, the "body" has been forgotten. There was a grand decree, at the height of the French Revolution, that it be enshrined in the great temple of secularism, the Pantheon, but the Reign of Terror and the rivers of blood that accompanied it washed away most thoughts of actually carrying it out.

But not entirely all. In 1927—with Paris deep in its syncopated Jazz Age, Lost Generation sway, with Hemingway and Fitzgerald and Picasso and Stravinsky giving the city and the world a new way to understand "modern"—a couple of government officials happened to notice that the revolutionary decree to pantheonize Descartes had never been carried out. The pair brought the matter before the office of the consul general, an investigation was held, interest was piqued . . . and then a problem was discovered. Once again, the remains of the philosopher, now 277 years old, became international news. "The proposal to transfer the ashes of René Descartes, the seventeenth century philosopher, to the Pantheon, which has just been made by two general councillors, M. Robert Bos and M. André Gayot, may encounter a number of objections, according to the French Press, chief of which is the uncertainty as to the present resting place of the remains." So began the story that ran in the *New York Herald*.

"Where are the remains of Descartes?" *Le Temps* asked. "A controversy ensued yesterday . . . not on the doctrine of the philosopher but on the place, the state, and the authenticity of his remains." The story went on to note drily: "A plaque affixed to the church of Saint Germain-des-Prés indicates that they repose there. A plaque is not, one must say, irrefutable proof."

Government officials had unraveled the whole chain of historical events and personalities, from Chanut and Christina to Paul Richer and his exotic tripartite bust, and they came to conclude that there was a weak link somewhere. Why, when the coffin was opened before the final burial, did it contain mostly bone chips and powder? These were not proper remains. Digging up the coffin would be unlikely to yield further clues since no one doubted the reports of Delambre and others who attended the 1819 reburial. A problem must have occurred earlier. Somewhere—over six countries, across three centuries, through three burials—something had gone wrong with the handling of the remains of René Descartes. Once this was realized, the matter was dropped—no one wanted to go to the expense and bother of digging into the ancient church for a few handfuls of dubious powder—and it has largely remained dropped since that time.

Of course, it doesn't matter. There are something like thirty-one wars being fought on the planet as I write. We are on the verge, or over the verge, of irreversible environmental damage. There are epic confrontations brewing that transcend national borders, which have to do with clashes of religious, economic, and political systems. The insignificance of some bits of bone lying beneath the floor of a church in Paris—whether or not they ever belonged to the person to whom they are attributed—is fairly monumental.

But my point in pursuing the trail of Descartes' bones has been metaphorical: uncannily, they seem to form a spine, if you will pardon the expression, of modernity itself. Sixteen years after Descartes' death, Hugues de Terlon considered Descartes someone who had penetrated to the mystical heart of nature, and he took a finger bone as a religious relic, a holy object meant to bridge the gap between the material and the eternal. By the time of the French Revolution, Condorcet and his compatriots looked on the bones in the mirror-opposite way: as relics of secularism, symbols of the force that reoriented people and society toward the here and now and gave rise to the principles of individual liberty, equality, and democracy. For Berzelius, Cuvier, and other nineteenth-century scientists, the skull was a talisman of science. Descartes' bones—or rather, the meanings people have attached to them—are really about who we were and are, including the convictions and confusions and confrontations that divide us.

Quite beyond that, a search for answers of the sort I've been conducting is itself utterly Cartesian, is it not? The Cartesian method lies beneath not only the scientific method but other modern modes of inquiry. We are a probing, analytical culture. Of course, a certain amount of baggage comes with this. The American philosopher John Dewey characterized the modern mind since Descartes as being caught up in a hopeless "quest for certainty"—hopeless because certainty does not exist in the real world. In absorbing Descartes' mind-body dualism, we have set up our understanding of knowledge this way: there is a fixed world of objects that is "out there," and there is a mind "in here." Knowledge is what happens when the mind reaches out to that fixed world. Dewey called this view a "spectator theory of knowledge." A thing is real, we think, if we can see it and hold it, if it can be associated with a particular time and place, if it happened in history. But according to current thinking in philosophy and science, this is not how reality works. We are supposed to understand that things that seem to be hard and clear and certain are in fact floating on a sea of probability. Contingency—whether in nuclear physics or in morality or in our personal relationships—is a governing principle. Like children outgrowing fantasies, we are supposed to realize ultimately that there is no such thing as certainty.

But we want it anyway. We probe the world and our past. Were the American founding fathers heroes or slave drivers? Who was Jesus and how do intelligent Christians square his supposed miracles with our understanding of the physical world? We are all part-time historians, genealogists, doctors, investigators. We demand information about our child's illness; we hire private detectives to check up on our spouse; we haul that piece of furniture that has been in the family for generations to an appraiser to learn its value; we have difficult conversations with parents about the circumstances of our birth. The archetypal modern figure may not be Descartes or Galileo or Einstein but Sherlock Holmes. The success of Arthur Conan Doyle's stories was in no small part due to their capturing the modern fascination with inquiry that was part of Victorian society—the idea of applying observation and analysis to the ordinary world and revealing the hidden structure of things. We are all detectives. Sifting clues and making deductions is in our blood—or perhaps better to say in our brains.

And we crave closure. What actually happened to the "body" of Descartes' bones is a mystery, but we like mysteries when they come paired with solutions. So here is mine.

I AM STANDING ON A wide plaza—the highest point in Paris—looking at one church but seeing two. I'm holding a postcard of an ink drawing of the scene before me as it existed four centuries ago. The real-life church in front of me, with its somberly busy façade mixing elements of Gothic and Renaissance architecture, is nearly exactly as it appears on the postcard. But where, on the postcard, immediately to the right of it—wall to wall, in fact—stands a church with a sedate Gothic brick façade, there is now a gap, a narrow street that connects this plaza, the place du Panthéon, to the street behind (the rue Descartes, no less) and the warren of Latin Quarter streets that descend to the Seine. By squinting and folding the postcard in half I can replace the missing church, relocating it up against the wall of the existing one.

The church on my right—the one that isn't there anymore— was Ste.-Geneviève. Here, on an early summer's evening in 1667, friends and followers of Descartes, having orchestrated the disinterment and transfer of his remains from Stockholm, laid them to rest on French soil, near the relics of the city's patron saint, Geneviève herself. Here the bones of Descartes remained for more than a century until, in 1792, anti-Catholic mobs threatened the church, so that its abbot begged the keeper of the revolutionary government's *dépôt des monuments* to come and save what could be saved.

Alexandre Lenoir—that sepulchrally dashing rescuer of artifacts—took official charge of the situation and went to work. He took sculpted figures, columns, plaques, and markers. He took the tabernacle. He took coffins and sarcophaguses. He took bones. But he did not take the remains of René Descartes. It was a time of chaos and violence, of roving gangs, makeshift roadblocks, burning buildings. For Lenoir, given his business, it was an intensely busy moment, as he sent his small team of assistants to retrieve objects from every part of the city. Speaking specifically of the moment when he supposedly retrieved the remains of Descartes from Ste.-Geneviève, Lenoir wrote four years later that "the circumstances of that time were not happy ones for me." We have no way of knowing what this personal source of unhappiness was, but his suggestion is that he was distracted.

Though he was trained as a painter, Lenoir's skill was not in art; he fancied himself a scholar of architecture, but he made basic mistakes in his descriptions of buildings and their features. However, as his biographer has said, Lenoir was a registrar of the first order: his records of his holdings are unparalleled, and an important database of French art and architecture; his renderings of the tombs in the church of Ste.-Geneviève are meticulous. And yet he made no record of recovering the remains of Descartes.

Decades later, long after Lenoir's beloved Museum of French Monuments was closed, and after Berzelius discovered the skull purported to be Descartes' in the collection of a casino operator in Stockholm, Georges Cuvier, seeking to unravel the tangle of threads surrounding Descartes' bones, wrote to Lenoir at his home in Paris. There was a considerable disparity between the two men: Lenoir was at this point more or less relegated to obscurity, while Cuvier, as permanent secretary of the academy, was one of France's great figures. The letter was written under the seal of the Academy of Sciences and carried the weight of authority:

I pray you, Monsieur, to be so kind as to tell me what you know regarding a fact that can contribute to confirm or deny the authenticity of the head that was sent latterly from Stockholm to Paris and that passed in Sweden as being that of Descartes.

We need to know if, while the remains of this philosopher were being carried to the Petits-Augustins [the convent on the Seine that Lenoir turned into the Museum of French Monuments], there was a head or some part of the head.

Monsieur Berzelius, who was in Paris during the burying of these remains at St.-Germain-des-Prés, heard it told by one of the persons who had been present at the ceremony that the head was not found there and that it was believed to remain in Sweden; Monsieur Delambre, who saw and examined these remains, assures also that there was no recognizable fragment of the head.

And yet Monsieur de Terlon, minister of France to Sweden, who occupied himself in 1666 with the return of this store to France, seems to have taken the greatest precautions to assure its integrity; he would had to have been deceived by the persons whom he had charged with the packing . . .

Cuvier's focus was on the skull, and his tone formally polite, but he was clearly suggesting some mishandling of the remains between the time Terlon packed them in Sweden and Delambre and others saw them. And the man most associated with them during that time—who had supposedly retrieved them from Ste.-Geneviève and then held them in his keeping for twenty-seven years—was Lenoir.

Lenoir wrote back at once. He didn't even bother to pick up a fresh sheet of paper but scrawled his reply on the back of Cuvier's letter:

Paris, the 16th of May 1821
To Monsieur le baron Cuvier
Monsieur le Baron,

I hasten to respond to the letter that you did the honor to write me, relative to the mortal remains of René Descartes. . . . Messieurs the abbé Saint-Léger, Le Blond, and I took ourselves to the church to make a search for the body of Descartes; and we dug the earth around the pillar to the right of the entrance, where a medallion in terra cotta was attached and where graven inscriptions on white marble marked the mausoleum of Descartes. At a very little depth in the earth, we found the remains of a coffin of porous wood and some very disappointing bits of bone in very small quantity, which is to say, a portion of a tibia, of a femur, and some fragments of a radius and a cubitus.

Please note, Monsieur, that these fragments, of which there would have been twice as many had the body been laid there in its entirety, were alone and isolated from other parts of the skeleton, which were missing . . .

He went on to say that he had found one small piece of bone plate that could have been a skull fragment; it was this that he had cut into circles and distributed to friends as rings.

But two years before Cuvier's letter Lenoir had also been asked about the transfer of Descartes' remains, and on that earlier occasion he had given a somewhat different answer. In the immediate aftermath of the reburial in the church of St.-Germain-des-Prés, and perhaps in response to questions from Delambre, the city's conservator of public monuments had wanted to look at the records from the time when Lenoir dug up the bones, but he couldn't locate any reference. "You will not have found, Monsieur, the record of the exhumation of Descartes in the archives of the museum," Lenoir informed him, "because none was written up." The reason, Lenoir said, was that "this operation was conducted revolutionarily"—that is, during the chaos of the Revolution. It took place, he went on, "in the presence of the police commissioner of that section of the *quartier*" and "at the request of Messieurs the abbé Saint-Léger and Le Blond," who were both members of the Monuments Commission. "They are both dead now," he added, then noted that "the remains . . . were brought to the Petits-Augustins by a commissioner whom I paid."

Lenoir appears to be suggesting that he himself was not present when the remains were dug up. "At their arrival at the museum, I placed [the remains] in an antique porphyry tomb," he continues, painting a picture of others as having not only dug in the church but carried what they found to the grounds of the museum, where he received the remains and put them into the ancient sarcophagus. The seemingly superfluous note that Saint-Léger and Le Blond had both since died—and were thus not available to corroborate his account—also sounds like a reflexive covering up of an awkward or embarrassing fact.

The awkward fact, it seems to me, is that Lenoir failed to get the remains of Descartes. It was an unhappy time, there was the chaos of the Revolution, he sent others—as indeed he often did. Then there is the evidence. Lenoir says that what he—or his assistants—found was "the remains of a coffin of porous wood" and some meager fragments. But there was no wooden coffin. The burial ceremony of 1667 had been a grand affair—recall that the Cartesians had pulled every string in order to have the translation and reburial of their great hero treated as an officially sanctioned event—and the copper coffin in which Terlon had originally packed the remains was accompanied by a copper sword on which Claude Clerselier, the editor of Descartes' papers, inscribed an account of the translation and the names of those taking part. This work was done, in the words of Baillet, "in the presence of these friends."

So where was the copper coffin? Where was the copper sword? What's more, Lenoir had believed that one of the larger bone fragments—from which he had made his rings—was a piece of skull. A portion of skull is precisely what should not have been found, the skull being the one piece of the skeleton that has the best documentary reason for not being among the remains.

But the most decisive point against Lenoir is found in his letter to Cuvier. He says that he and two others "took ourselves to the church to make a search for the body of Descartes; and we dug the earth around the pillar to the right of the entrance, where a medallion in terra cotta was attached." Such a plaque to the memory of Descartes may well have been placed on a pillar to the right of the church entrance, but that was not the location of the tomb. Records of those who participated in the burial at Ste.-Geneviève are clear in stating that when the prayers and songs were finished the coffin was carried "to the southern end of the nave and put against the great wall, between two confessionals, into a vault that had been intended for it, between the chapel of the titular Saint Geneviève and that of Saint François." Descartes' resting place was not the spot under the floor of the church where Lenoir and/or his assistants dug but a vault along the southern wall.

All of this points to Lenoir as the man responsible for breaking the historical chain—for losing the bones. He must at first have thought that he had gotten the genuine remains of Descartes from the church, for we know how he cherished the remains of great men and women as secular relics, and he went to the trouble of turning these particular bones into jewelry. But in time he seems to have come to realize what had happened, and begun making excuses. The remains—presumably a more or less complete skeleton with the exception of its skull—would probably have been dealt with by revolutionary ransackers. If the vandals had somehow overlooked them, Descartes' bones would have been churned up when the ruined church was finally destroyed in 1807 to make way for the road.

The Cartesian tendency of favoring mind over matter—mind over body—thus has a metaphorical cap. The skull—the representation of mind—having been subjected to repeated and increasingly sophisticated scientific study and judged to be authentic, sits enshrined in a science museum, the Musée de l'homme, or Museum of Man, which was formed in 1937 from older collections of anthropological artifacts. Indeed, as I write, it is part of a special exhibition at the Musée de l'homme entitled *Man Exposed,* sitting beside a Cro-Magnon skull to demonstrate the breadth of human thought and accomplishment over the millennia, once again serving as the very representation of "modern." As for the body, the trail ends abruptly, veering sharply into oblivion. And that is perhaps as it should be. Dust to dust. *In secula seculorum.*

7

A Modern Face

OMETIME AROUND 1985, A JAPANESE TELECOM-munications engineer named Hiroshi Harashima was trying to develop a practical videophone when his thinking suddenly jumped from the technical to the philosophical. He describes his normal work as "how to connect terminals such as a telephone set," but, he came to wonder, "are not the true terminals human beings rather than telephone sets?" The prime difficulty researchers had found with videophones was not technical. It had to do with the human face. Faces, Harashima realized, convey emotional rather than practical information. Researchers had discovered that people were used to the narrower emotional space of telephone communication and didn't feel comfortable with the deeper level of opening up involved in showing their faces. As Harashima explored the implications of this finding, his usual network of colleagues expanded. He was soon in touch with psychologists, forensic scientists, makeup artists, anthropologists, even mask collectors. He formulated the idea that everything he had formerly been involved in was contributing to the development of what he called "line humanity," a society built around streamlined high-tech communications that flattened information. The face, in contrast, was the oldest communication technology yet a deeply sophisticated one in which we expose a range of meanings, including many—making promises, asking for trust—that knit the culture together. "It is not too much to say that [the face] is the basis of social order," he wrote. He perceived dangers in this faceless society that was developing.

In 1995 Harashima and Hisao Baba, an anthropologist and anatomist who was also a curator at the National Science Museum in Tokyo, cofounded the Japan Academy of Facial Studies. Their intention was to bring together experts in fields as diverse as orthodontics, cosmetology, biology, dentistry, and drama in an evolving project to explore ways to keep one of humanity's most basic communication technologies relevant in a digital world. Projects associated with the Academy of Facial Studies have analyzed wrinkles and aging, studied how people remember facial expressions, and described brain responses to emoticons.

In 1999, the academy and the museum intended to hold a vast exposition entitled the Great Exhibition of the Face. It so happened that before the exposition was planned Baba was in Paris visiting the world-renowned collection of Neanderthal skulls at the Musée de l'homme. While showing him around, Philippe Mennecier, the museum's director of conservation, happened to mention that the museum possessed the skull of Descartes. Baba was "thrilled" to behold the curious object, and as the plans for the exhibition on the face were being drawn up, he had an idea.

Some months later, Mennecier received at the museum a representative of the firm of André Chenue, a company that began its life in 1760 as transporter of the personal belongings of Marie Antoinette and is today one of Europe's leading shippers of fine art. The representative carried a custom-built box—of smooth-grained blond wood with elegant hasps—that was not much larger than a human head. When the representative left the museum, still carrying the box—now packed with cargo—he was in the company of André Langaney, the director of the museum's laboratory of biological anthropology. Later, Langaney boarded a flight for Tokyo and took a seat in business class. His carry-on luggage was the box. Memories of the debacle during the flood of 1910 died hard; the agreement to send the skull of Descartes to Tokyo—the first time it had been out of the museum since 1912, when it spent two hours on display at the Academy of Sciences—came with the condition that it be treated as an object of inestimable value. "It seemed to us out of the question to leave the skull in the luggage hold," Mennecier told me.

At the time, Baba knew nothing of the skull's history. His interest was not in authentication. The exhibition in Tokyo would feature images of fashion models and Papua New Guinea tribesmen, dental surgery demonstrations, Rembrandt portraits, Noh theater masks, computer-assisted facial reconstructions, Abraham Lincoln's death mask, cartoon caricatures, tapes of Japanese actresses of the 1950s, and Photoshopped Mona Lisas with frowns and grins. It was to be a compendium of facial data, but also a demonstration of human technologies and of one of the current trends in science, the interdisciplinary approach. Baba's idea was that the exhibit should have an introductory piece that set up its historical and intellectual features. He himself was an anatomist who had performed many autopsies. He wanted to apply his skills to the skull of Descartes to give the facial exhibition a face. The skull would be both an introduction to the exposition and, as he said, "a symbol of homo sapiens, a symbol of human intelligence, the symbolic face of modern man."

The first step was to perform a thorough forensic analysis of the skull. The teeth—in particular the wisdom teeth—indicated that it was the skull of an adult. Evidence of "rough" attachment of the temporalis muscle suggested a person of middle age. Baba's analysis would have pleased Louis-Pierre Gratiolet, who sparred with Pierre-Paul Broca over the matter of cranial capacity, for while Baba noted that "the braincase is wider than that of the average European male today," he also concluded that "it is almost certain that the bones and muscles of the neck were smaller than those of the average European male, and Descartes was known to be a small man."

If there was any new information that came out of Baba's study, it concerned diet. The teeth and muscles of the jaw indicated that the subject "had healthy teeth and ate plain, rough food." More broadly, Dr. Baba concluded that "the individual who had this skull was a small and slender late-middle-aged European male. He is highly likely to have been a man of the late medieval or early modern era."

Working with a sculptor, Baba made a plaster cast of the skull, then applied muscle, cartilage, and skin and compared the head to what he took to be the most accurate portrait from life: the one in the Louvre. "The shape of the bones reflects the person's facial appearance quite precisely," he wrote. "This skull shows similar characteristics to the portrait of Descartes. The head was restored by adding muscle and skin onto the replica of the skull and there was no difference between the restored head and the portrait of Descartes."

From this point, the anthropologist and the sculptor worked from the painting to create a life-sized bust. This, side by side with the skull, formed the introduction to the Great Exhibition of the Face. It opened on July 31, 1999, ran through May of 2000, and had 300,000 visitors.

Interestingly, Baba told me he had no idea of Paul Richer's work. Without even being aware of it, he had duplicated Richer's experiment, right down to comparing the results against the Louvre portrait. At the same time, Baba brought Descartes' bones into the era of computer analysis and interdisciplinary study. Like Richer, Baba used the skull to fashion a life-sized bust, giving a face to the bones. The bust created by the Japanese anthropologist and artist is less bold and dashing than Richer's; it has instead a simplicity and calmness. Baba gave a copy of it to Philippe Mennecier, who showed it to me where it is kept in a basement studio at the Musée de l'homme, remarking, with his dry delivery (and, I thought, some truth): "It's a good likeness, though I find the eyes look rather Japanese."

I F IT'S FAIR to assign the credit or blame for modernity and its problems to any one person, Descartes is the prime candidate. Richard Watson, the American philosopher and biographer of Descartes, considers him sweepingly elemental:

> Descartes laid the foundations for the dominance of reason in science and human affairs. He desacralized nature and set the individual human being above church and state. Without Cartesian individualism, we would have no democracy. Without the Cartesian method of analyzing material things into their primary elements, we would never have developed the atom bomb. The seventeenth-century rise of Modern Science, the eighteenth-century Enlightenment, the nineteenth-century Industrial Revolution, your twentieth-century personal computer, and the twenty-first-century deciphering of the brain— all Cartesian. The modern world is Cartesian to the core.

Of course, along with all of these achievements came the political strife and the clash of worldviews that dominate our times.

Between them, with their various methods and working at opposite ends of the twentieth century, Paul Richer and Hisao Baba managed to give Descartes a human face. As Baba said, his ambition was to work outward from the skull to create a "symbolic face of modern man." What, though, in the early twenty-first century, is the face of modern humanity? What are its features? In what direction is it looking? Those are the questions that drove me in the first place to investigate the story of Descartes' bones.

One bright winter day in 2007, I found myself in the restaurant of a fashionable New York hotel having lunch with Ayaan Hirsi Ali, the internationally famous Somali-born promoter of the rights of Muslim women and one of the most admired and reviled women of the early twenty-first century. I was interviewing her for a magazine article, and as she talked—about what she sees as the dangerous irrationality inherent in all religion, but especially in Islam, and about how she believes the West contained its own irrationality by cordoning off faith from politics—it occurred to me that hers is one of the faces of modernity; her life story encompasses the legacy of Descartes' bones. As a small child, she sat with her grandmother under a talal tree, shaded from the desert sun of Somalia. At five she was held down on a kitchen table while a man who was probably "an itinerant traditional circumciser from the blacksmith clan" snipped out her clitoris and inner labia with a pair of scissors. She grew up against a background of resurgent Islam, eventually fled to Europe, asked for asylum in Amsterdam, and began what she called "my freedom." She enrolled at Leiden University (where Descartes had spent time while awaiting the publication, by a Leiden printer, of the *Discourse on the Method*), started reading Voltaire, Rousseau, Marx, and Freud, and turned her back on her faith and tradition. "Meeting Freud," she wrote in her 2006 autobiography with declarative understatement, "put me in contact with an alternative moral system."

From religious fundamentalism she veered toward what some of her critics have called Enlightenment fundamentalism. She became a member of the Dutch parliament, and from that platform gained attention as a critic of religion in general and of Islam in particular. When a young Amsterdam Muslim murdered Theo Van Gogh, the filmmaker with whom she had collaborated on an anti-Muslim film, and pinned a note to his chest threatening her life as well, Hirsi Ali became a global phenomenon; as she moved from Europe to the United States, she used the fame to amplify her views on immigration, cultural identity, and the interaction between the Muslim world and the secular West.

Hirsi Ali's beliefs are as sharp as a knife blade. Reason is the great light of humanity; religion is a force of chaos and darkness. "The West," she told me, "was saved by the fact that it succeeded in separating faith and reason. That led to secular government. Secular government is built on human reason, with all its fallibility. Faith assumes infallibility, and that is the danger. Our prophet Mohammad can never make a mistake, so we are stuck with him." The fault line Hirsi Ali stands on runs back through the centuries of the modern era; the tensions that swirl constantly around her (as we ate lunch her round-the-clock security team hovered nearby) harken back to the forces that squared off at Utrecht University in the 1630s, when Regius, Descartes' first disciple, presented a form of Cartesianism to an awestruck public, and to those that fueled the French Revolution. Only today those tensions are global. The challenge that Islam poses, in the view of many, is one of a culture that has not experienced the centuries of modernity—that has not lived with Descartes' bones.

By the time I sat down with Hirsi Ali I had been working on this book for two years, and I was used to discovering resonances between its themes and current events. While writing a cover story for the *New York Times Magazine* about Pope Benedict XVI and his efforts to renew the Catholic Church in Western Europe, I realized that everyone I met and interviewed—clergy, lay Catholics, European Muslims, all of them roiled by the tensions between secular Europe, Christianity, and Islam—was living on this fault line of modernity.

On another magazine assignment, I found myself in the living room of a ranch house in suburban Maryland, sharing a meal with six people who were grassroots organizers of a movement in that state to amend its constitution to forbid same-sex marriage. Their feelings stemmed from their religious beliefs about homosexuality: it is a sin and a disease; it has no reality in the individual but is rather a sickness in society. That the World Health Organization, the American Medical Association, and virtually every other professional medical and psychological organization takes a completely different view only reinforces the convictions of these people, for they see those organizations as based on shaky, non-biblical foundations. There have always been people at the fringes of Western society who refused to go along with the basic ideals inherited from the Enlightenment—views about the individual, the primacy of reason, and so on. But the beliefs I encountered in Maryland extend far and wide in America, where Christian absolutism is a major force. They go beyond human sexuality to biotechnology, education, social services, child development, and virtually every facet of life. They have influenced the foreign policy of the world's only superpower. In this system of beliefs modern history turns out to be a series of wrong turns. The group in Maryland pointed these out for me as neatly as if they were holding bulleted outlines. The women's rights movement. The birth control pill. The idea of a separation between church and state. Darwin. Finally, one man, a minister, said the name that was sitting in the front of my mind: "If you think about it, it really all starts with Descartes." He then went on to talk eloquently about the changes that began with Descartes' reorientation of reality around human reason. "The human mind can be led astray," he said. "It is no basis for anything without God."

The historical nature of these clashes of worldviews is striking enough that it has become part of the public discourse. Each generation interprets the past according to its own needs, and lately—as people have found themselves faced with such challenges to modern secular society—the Enlightenment has come back in vogue. In the twentieth century, it was first reanimated in the World War II era. A handful of scholars who came of age under the threat of Nazi domination used the history of the eighteenth century and the work of the great Enlightenment thinkers as a beacon to light the darkness they themselves were living through. Ernst Cassirer's *The Philosophy of the Enlightenment*, which appeared in 1932, was intensely read throughout the war and in the postwar period and was instrumental in creating the twentieth-century view of the late eighteenth century as a time of reason beating back the forces of chaos. Cassirer was a German Jew who emigrated to the United States during the war; his objective was to use history as a tool. He cast his work as a frank appeal: "The age which venerated reason and science as man's highest faculty cannot and must not be lost even for us. We must find a way not only to see that age in its own shape but to release again those original forces which brought forth and molded this shape."

At about the same time, the American historian Carl Becker

gave a shorthand account of the beliefs that the twentieth century had inherited from the early modern era: "(1) man is not natively depraved; (2) the end of life is life itself, the good life on earth instead of the beatific life after death; (3) man is capable, guided solely by the light of reason and experience, of perfecting the good life on earth; and (4) the first and essential condition of the good life on earth is the freeing of men's minds from the bonds of ignorance and superstition, and of their bodies from the arbitrary oppression of the constituted social authorities."

This is the modern creed—or it was until a generation ago. Things changed in the 1960s and 1970s. After the Vietnam War and the social upheaval of that era, such a grandiose view of history began to seem precious and stale. Were the men and women of the eighteenth century really such paragons? Does history ever actually work that way? Is there really a march of progress, with each generation building on the work of the last and moving forward toward some ever-brighter future? If modern Western history was such a grand parade, how did one account for colonialism, Nazism, Soviet-style communism, for slavery and gulags and concentration camps? Postmodernism replaced progress with skepticism.

Then a new millennium—to be precise, September 11, 2001— brought a sudden turn of thinking, and a reappraisal. The threat from some quarters has seemed bewilderingly ancient, as if a dinosaur had suddenly reared up from its prehistoric slumber. One statement of it came in the form of a letter written in 2006 by Iranian president Mahmoud Ahmadinejad to American president George W. Bush: "Liberalism and Western-style democracy have not been able to help realize the ideals of humanity. Today, these two concepts have failed. Those with insight can already hear the sounds of the shattering and fall of the ideology and thoughts

of the liberal democratic systems." Instead of democracy, Ahmadinejad predicted, "the will of God will prevail over all things."

Part of the inheritance of modernity has been the idea that its
core values of democracy and individual liberty have the force of
inevitability. They emerged at a given point in history, and now
that they have arrived we tend to think that everyone acknowledges them as universal values. But that doesn't seem to be the
case. José Casanova, a sociologist of religion at the New School
for Social Research, told me that the idea used to be that Western
development "was prescriptive for the rest of the world, that it
would be a model for other societies, so that these other societies
would follow a secular path. But now throughout the world you
find religious revivals. We're learning that more modern societies
don't necessarily become more secular." And whether in Western
or non-Western tradition, theocracy has tended not to sit easily
alongside concepts of democracy and individual freedom.

The recognition of the sobering fact that these modern ideals
are not necessarily spreading around the world—that, possibly,
they are not inevitable at all but could rather be fragile, ephemeral, temporary themes in world history—has coincided with a
desire to look back at our past to remind ourselves of what we
are. I think that is good and necessary; I agree, for example, with
the German scholar Heinz Schlaffer when he says that "Western
culture is also fundamentalist: Its fundament is called the Enlightenment" and that "the paradox is that this fundament is the
basis for our present society, but also half forgotten by it." Hirsi
Ali formulated her message for me in this way: "The only way to
stand up to radical Islam is to revive the Enlightenment, the message of the Enlightenment, and make the people who inherited
all of this"—and here she waved her hand toward the window and
the skyscrapers of Manhattan—"realize that this all just didn't

fall out of the sky. There is a long history of struggle behind the development of this society. And religion, including Christianity, has most of the time hindered that development."

I think this is largely true—and I think in its idiosyncratic way the history of Descartes' bones sketches the long journey, filled with false starts and blind alleys, that led to modern society—but as it narrows, this line of thinking darkens. I suspect that much of the talk about valuing the Western tradition is cover for a brutish us-or-them impulse. In these pages I have taken up Jonathan Israel's thesis that there was a three-way division that came into being as modernity matured. There was the theological camp, which held on to a worldview grounded in religious tradition; the "radical Enlightenment" camp, which, in the advent of the "new philosophy," wanted to overthrow the old order, with its centers of power in the church and the monarchy, and replace it with a society ruled by democracy and science; and the moderate Enlightenment camp, which subdivided into many factions but which basically took a middle position, arguing that the scientific and religious worldviews aren't truly inconsistent but that perceived conflicts have to be sorted out. All three of these factions remain with us today. Their adherents express themselves on TV news talk shows, in blogs and opinion pieces, and in court cases. Those who promote "intelligent design" as a replacement for the theory of evolution are members of today's version of the "theological party" who are attempting to infiltrate the moderate camp. In his best seller *God Is Not Great,* Christopher Hitchens sounds the trumpet for "radical Enlightenment" warriors of the twenty-first century by using language that mirrors the freethinkers of three centuries ago: "We distrust anything that contradicts science or outrages reason. . . . The person who is certain, and who claims divine warrant for his certainty, belongs now to the infancy of our species."

Hirsi Ali "converted" not just to secularism but to its radical form. She would have found a ready place for her ideas during the French Revolution, and indeed her ideas tend toward a similar extreme: she has declared that "we are at war with Islam" and that, in the name of reason, not just Islamic terrorism but Islam itself, along with its 1.5 billion adherents, must be "defeated" so that "it can mutate into something peaceful."

This is patently frightening talk, and I believe it exposes the flaws of radical secularism. I agree with the radical secularists that enormous ugliness has been done and is being done in the name of religion, and I think that we have to find an intelligent way not to tolerate religious intolerance, but I believe history shows that there is lethal error in radical secularism—or rather, two errors. First, it thinks too highly of reason, or of the ability of humans to employ it. The history of modernity, even the anecdotal version of it that comprises the story of Descartes' bones—scientific stupidities proliferating majestically alongside real advances—makes plain that trying to follow reason is not the same thing as being right, and every successful demagogue of the twentieth century has demonstrated how easy it is to manipulate reason and direct its course from the truth to something like its opposite.

One could argue that the antidote is to recognize the tendency to misapply reason and try to correct for it. But that ignores the second error, which is the greater. In the interest of pursuing its own brand of certainty, radical secularism takes a too narrow construction of reality. It puts on blinders. Religion, like art, is a way of negotiating the complexity that the philosophical puzzle of dualism, and all of the attempts at overcoming it, acknowledges.

To deny religion outright exposes those who trumpet the use of reason to the charge of unreasonableness—of intolerance.

If there is a solution to the dilemma of modernity, surely it lies in bringing the two wings into the middle, which is where most people live. Jürgen Habermas, the great German philosopher (who himself is not religiously inclined), has used the term *postsecular* to describe what he believes can be the next stage in the evolution of Western society. This stage, he argues, would involve "the assimilation and reflexive transformation of both religious and secular mentalities."

Such transformation would presumably require convincing or teaching or cajoling or arm-twisting the radical partisans—the theologians and the radical secularists—into widening their picture of reality, getting both to acknowledge that they don't have a lock on truth, that the world is too wild for our strategies to contain it. At the same time, it surely would mean finding a way to convince one of those wings—the billions of people who grew up in cultures without the legacy of Descartes' bones—to recognize that we have, in the past few centuries, latched onto some fairly comprehensive ways of understanding the world and advancing humanity and that these must be taken by everyone as a foundation.

The task (which may be impossible to achieve, but is there any alternative but to try?) might be translated into Hiroshi Harashima's terms: to move away from "line humanity" and put faith in the oldest of communication technologies. Then, maybe, members of opposing camps could meet anew, seeking out signs of trust in another person's face.

Epilogue

O N FEBRUARY 11, 2000, EXACTLY 350 YEARS AFTER the cold Stockholm night when René Descartes breathed his last, a group of about twenty men and women gathered in the stony chill of the church of St.-Germain-des-Prés in Paris at a Mass for the eternal rest of the philosopher's soul. The priest performing the service was Father Jean-Robert Armogathe. Of all the "amoureux de Descartes," as Philippe Mennecier, the current keeper of the skull of Descartes, refers to the small coterie of people whose interest in the philosopher carries over to his mortal remains, Father Armogathe was the one I saved for last to contact, not because I thought he would make a fitting end for this book but out of intimidation. Besides being a high official in the diocese of Paris who was once the chaplain of Notre-Dame cathedral, Armogathe is one of the preeminent Descartes scholars in the world, director of studies of the History of Religious and Scientific Ideas in Modern Europe program of the Ecole Pratique des Hautes Etudes at the Sorbonne, author of works on Cartesian science, on correlations between the Bible and seventeenth-century science, and on that issue at the heart of the Cartesian crisis of dualism, the transubstantiation of the Catholic host.

Others who helped me in my research—themselves prominent philosophers—said, when I inquired about him, that yes, indeed, I would be remiss if I did not seek out Père Armogathe. But they warned that he could be a forbidding presence. "Don't be surprised if you don't hear back," said Richard Watson, probably the foremost American scholar of Descartes. Eventually I e-mailed the French philosopher-priest, telling him of my project, saying that I planned to be in Paris, and wondering about the possibility of meeting. Surprisingly, Armogathe replied at once, saying he was willing to talk; even more surprising was the way he closed his e-mail—"Looking forward to meet u!"—which certainly stripped away some of the veneer of austerity.

A few weeks later I stood before an ugly 1970s-era building directly opposite the Luxembourg Gardens, the headquarters of the Institut Bossuet, the Catholic boarding school of which Armogathe was the rector. I was shown into his ground-floor office. The windows looked across at the park, with its high iron gates topped with gold spearheads, through which I could see a pair of lovers on a bench and a class of art students, pads in hand, grouped around a statue of twisting Baroque figures. I studied the office while I waited. The desk was piled with stacks of papers. On the walls were a framed engraving of Descartes and one of Baillet, Descartes' seventeenth-century biographer. On a bookshelf, propped against the complete works of Galileo in Latin, were snapshots of Armogathe with Pope John Paul II.

Suddenly the man himself burst in. He was small, stoutish, compact, gray-haired, almost comically vigorous for such a serious person. He was in constant motion, darting or lunging. I asked about his past and he told me a story that seemed right out of *The Da Vinci Code*. "In the 1980s when I was chaplain of Notre Dame I actually had my lodging there—in the cathedral itself," he said. "In the nineteenth century they had built lodgings for the custodian of the cathedral up in a kind of loft, which were reached by a pseudomedieval corkscrew staircase. It was the most dramatic apartment you could imagine. I had this huge dining room looking over the Seine on three sides. My kitchen window opened out onto the southern rose, where I had my own private terrace. Now the apartment has been destroyed, but I lived there for five years, and I loved it." Armogathe took to hosting barbecues on the terrace, and it became a thing in 1980s Paris to get an invitation to the soirees of the chaplain of Notre-Dame.

He changed the topic and talked about his work. For five years, he told me, he had been studying vision and optics from the seventeenth century to the present day. His argument was that from its beginning science took its ideas about vision from medieval and Renaissance Catholic notions of spiritual visions and inner light. The thesis he was formulating was that our scientific understanding of the sense of vision is built around spiritual metaphors. "I'm against the idea that there is a clear cut between the Renaissance and the modern ages," he said. "I think modern thinking gets its patterns from the theological realm. Biblical concepts allowed science to progress."

From 1996 to 2000, Armogathe had been pastor of the church of St.-Germain-des-Prés, the site of the last burial of Descartes. During that time, as the 350th anniversary of Descartes' death approached, he had had the idea to conduct a special Mass. What had motivated him to do so? I wanted to know about the event in itself, but really I was interested in a deeper point. Armogathe was both a priest and an authority on Cartesian science. It occurred to me that if anyone alive had insights into bridging the modern divide between body and soul—if anyone alive had perspective on both current science and contemporary spiritual concerns as they relate back to the birth of modernity and the so-called father of modernity himself—it must be him.

"In Catholic tradition," Armogathe said, "it is not only prayer for the soul of the departed. There is also the belief in the resurrection of the body. For Catholics, mortal remains mean something. Churchyards are not urban repositories for garbage but places of sleep and waiting. It's like there are seeds under the ground, waiting for spring to come."

As was the case during the Middle Ages, the heyday of relics, the Catholic Church still places special value on human remains. There is nothing surprising in the fact that Catholicism—or any other spiritual tradition—has the mind-body problem sorted out to its satisfaction. Somehow, faith dissolves dualism, unites body and soul.

But a theological answer wasn't satisfying. I posed the question differently, asking specifically about Descartes as the father of dualism and the modern mind-body problem. Like one of the first-generation Cartesians in late-seventeenth-century Paris who brought Descartes' bones from Stockholm, Armogathe was ready to defend the master. He disputed the idea that Descartes had invented dualism. Expressions of mind and body existing in different realms, he reminded me, go back to the ancient Greeks. This was a good point, and it could be taken further still. A word like dualism suggests an abstract puzzle, but it is grounded in the everyday. We are all philosophers because our condition demands it. We live every moment in a universe of seemingly eternal thoughts and ideas, yet simultaneously in the constantly churning and decaying world of our bodies and their humble situations. We are graced with a godlike ability to transcend time and space in our minds but are chained to death. The result is a nagging need to find meaning. This is where the esoteric "mind-body problem" of philosophy professors becomes meaningful to us all, where it translates into tears and laughter.

Dualism is thus in a sense universal, and philosophers have puzzled over it forever; yet Descartes' statement of it, I suggested to Armogathe, is the one that stirred a historic commotion in the Western world. His dualism is the legacy we live with.

In response, Armogathe pointed me in a new direction—toward what is perhaps the most meaningful "solution" to this puzzle for all of us. Late in life, he noted, Descartes tried to deal with the separation of mind and body. "In his last book, Descartes states that in effect there must be a third substance, which is not really a third substance but a compound of both mind and body," Armogathe said. "I should treat it as a code, an encoding, which allows mind to react on body and body on mind." Philosophers have devoted endless hours of thought to the mind-body problem, but there is a real-world way in which we all transcend it. This "encoding" is one of the commonest parts of our lives, and also one of the most precious. Its importance was what Descartes hit on near the end of his life.

For a long time, scholars thought that after the death of his young daughter, Francine, who was born out of wedlock, Descartes parted company with the girl's mother, Helena Jans. It was assumed that his attachment to Helena, the only woman with whom we know him to have been intimate, was only through their child and that he cast her aside after the girl was gone. But a few years ago a Dutch historian named Jeroen van de Ven did some inspired archival digging. Descartes lived at at least twenty addresses in Holland, and sometime after Francine's death he moved to the coast, a place of dunes and lashing winds called

Egmond-Binnen. In the notary records of the city of Leiden, Van de Ven discovered a marriage contract, dated four years after Francine's death, between one Helena Jans and a man from Egmond named Jan Jansz van Wel. Could it be the same woman? A kind of dowry was required—one thousand guilders, a considerable sum—and in a separate record Van de Ven discovered that the man who provided it was Descartes. Clearly, the couple stayed linked for four years after their daughter's death. Social standing prevented their truly being together even if Descartes would have wished it; he did, however, feel responsible, and in the end he provided a future for her.

During this time, the philosopher was being attacked mercilessly by Dutch theologians who were outraged at the implications of his work. Finally, he had had enough. "As for the peace I had previously sought here, I foresee that from now on I may not get as much of that as I would like," he wrote. "A troop of theologians, followers of Scholastic philosophy, seem to have formed a league in an attempt to crush me by their slanders." He would soon accept the offer of Queen Christina and go off to Sweden. He would leave behind not only the fury over what his work had unleashed but the mother of his child.

At this same time he began what would be his final work, and it may not be a coincidence that the book was a treatise on "the passions of the soul." Descartes had long before realized that there was a difficulty with his division of reality into mind and body—the difficulty being to figure out how the two substances interacted—and he was now moved to work on that confounding problem. His conclusion was that there is a connective tissue between the two, or, as Armogathe put it to me, "an encoding." The seventeenth-century terminology for this encoding was "passion." We might call it heart. This became the subject of Descartes' final work. Heart, he decided, was the interface between mind and body. Love, joy, anguish, remorse: we experience these in both body and mind, and somehow, Descartes became convinced, these passions link our two selves. He thus anticipated another modern field—psychology—in concluding that emotional states are tied to physical health and also to, as he would put it, "the soul."

But it was to be a purely philosophical exercise. His child was dead; he had married off the woman with whom he had been intimate. He seems to have given up on the "passions" personally. For him there was nothing ahead but cold and ice and death.

Even as he went off to die, however, Descartes gave Helena Jans a future, and the records of the little village of Egmond paint an evolving picture: Helena and her husband, Jan, live their lives at the inn that his family operates; in time, Jan dies and Helena inherits the inn; she remarries, and she and her new husband have three sons. There is no room in the records for all the life that must have been packed into the years, but somewhere in its density—the bustling of the inn, clanking tankards of Dutch beer, pipe smoke, leers and tears, song and suffering—lay the solution to Descartes' puzzle, which each of us solves, if we are very lucky.

Acknowledgments

I WAS FORTUNATE ENOUGH to receive guidance from three of the world's leading authorities on Descartes and his philosophy: Jean-Robert Armogathe, director of studies of the History of Religious and Scientific Ideas in Modern Europe program at the Ecole Pratique des Hautes Etudes at the Sorbonne; Theo Verbeek, professor of philosophy at Utrecht University; and Richard Watson, professor of philosophy emeritus at Washington University in St. Louis. I thank them for their help with this book and at the same time disassociate them from any errors it might contain. I am also particularly grateful to Philippe Mennecier, director of conservation at the Musée de l'homme in Paris, for giving me private viewings of the skull of Descartes, for sharing with me his dossier of documents related to its history, and for putting me in contact with other seekers after Descartes' bones.

My thanks also to the following: Susanna Åkerman, Swedish historian and biographer of Queen Christina, for showing me around Christina's Stockholm; Jane Alison, for Latin translations; Hisao Baba, director of anthropology, National Science Museum, Tokyo, Japan; Michael Baur, professor of philosophy,

Fordham University, for philosophical guidance and advice; Erik-Jan Bos, Utrecht University, coeditor of the new critical edition of the correspondence of Descartes, for his incisive critique of the manuscript; Friso Broeksma, for putting me up at his house on Prinseneiland; Veronica Buckley, biographer of Queen Christina; Jennifer Carter, McGill University, for her generous help with documents related to the Museum of French Monuments and for her thoughts on the quixotic figure of Alexandre Lenoir; Bernard Cartier, medical doctor and historian of French medicine; Hampus Cinthio, Historiska Museet, Lund, Sweden; Anne Deneufbourg, Maison Descartes, Amsterdam; I. M. L. Donaldson, emeritus professor of neurophysiology at the University of Edinburgh; Theodor Ebert, professor emeritus of the history of philosophy at the Universität Erlangen-Nürnberg, for sharing the manuscript of his monograph on Descartes' death with me; Marnie Henricksson, who had a great deal to do with the creation of this book; Suzanne Hoeben, for fact-checking; Beth Johnson and Jan Kat, for the loan of their beach house in Egmond aan Zee, which gave me a chance to steep in a land- and seascape where Descartes lived; Mami Kamei, for Japanese translations; Michael Martin for reading and critiquing the manuscript; Ann-Britt Qvarfordt, for Swedish translations; and Dominique Terlon, genealogist, for providing me with information on her seventeenth-century ancestor who carried Descartes' bones from Sweden to France and kept a piece for himself.

Thanks also to Armelle Aymonin, American Library in Paris; Marijke Bartels and Martijn David of Mouria Publishers, Amsterdam; Philippe Comar, Ecole des Beaux-Arts, Paris; Melissa Ann Danaczko; Maite Frost; Charles Gehring; Cobie Ivens; Jaap Jacobs; Raymond Jonas, professor of history, University of Washington; Birgitta Lindholm, keeper of manuscripts, Lund University Library; Therese Oltramari, librarian, Bibliothèque nationale de France; Emilie Stewart; Pamela Twigg; Theresa Vann, Hill Museum and Manuscript Library, St. John's University, Collegeville, Minnesota; Ulla von Wowern, Lund University Historical Museum, Lund, Sweden; Caroline van Gelderen; and Charles Wendell.

I would also like to acknowledge the main institutions where I carried out my research: the Bibliothèque nationale de France, the library of the Universiteit van Amsterdam, the New York Public Library, the American Library in Paris, the Bibliothèque Ste.-Geneviève, the Lund University library in Sweden, and Maison Descartes in Amsterdam.

As always, the greatest thanks to my agent and friend Anne Edelstein and to my brilliant editor, Bill Thomas.

Illustration Credits

Page 1 (top):   Musée de l'homme, Paris

Page 4 (bottom):   Photo by author

Page 5 (bottom):   Alexandre Lenoir (1761–1839) opposing the destruction of the royal tombs of the French monarchy in the Church of Saint-Denis, 1793 (pen & ink and wash on paper) (b/w photo) by French School (eighteenth century) © Louvre, Paris, France/Giraudon/The Bridgeman Art Library Nationality/copyright status: French

Page 7 (top):   Portrait of René Descartes (1596–1650) by Frans Hals (c. 1582/83–1666)/SMK Foto/Statens Museum for Kunst, Copenhagen.

Page 7 (middle):   Ecole Nationale Supérieure des Beaux-Arts, Paris.

Page 7 (bottom):   Portrait of René Descartes (1596–1650) c. 1649 (oil on canvas) by Frans Hals (1582/83–1666) (after) © Louvre, Paris, France/Lauros/Giraudon/The Bridgeman Art Library Nationality/copyright status: Dutch

Page 8 (top):   *New York Times*

Page 8 (bottom):   Japanese National Science Museum, Tokyo

Notes

Preface

xix   "You have a triangle": *Guardian,* February 26, 2007.

Chapter 1 The Man Who Died

3   "He who saves someone": Adam and Tannery, *Oeuvres,* vol. 5, pp. 477–78. Translated for the author by Jane Alison.

4   "devote what time I may still have": Descartes, *Discourse,* p. 96.

4   "The preservation of health": Descartes, *Philosophical Writings,* vol. 3, p. 275.

5   The life expectancy of a child: Data from chapter 5 of Clark, *A Farewell to Alms.*

8   "most men die of their remedies": Molière, *The Imaginary Invalid.*

8   "What says the doctor": The page's reply, meanwhile, seems to mock the practice: "He said, sir, the water itself was a good healthy water; but, for the party that owed it, he might have more diseases than he knew for."

8   "*dote not* upon, nor *trust*": Sym, *Lifes Preservative Against Self-Killing,* quoted in Andrew Wear, "Puritan Perceptions of Illness in Seventeenth-Century England," in Porter, *Patients and Practitioners,* p. 80.

11   the flexor digitorum superficialis: Masquelet, "Rembrandt's *Anatomy Lesson.*"

13   "hidden behind the scene so as to listen": Roth, *Descartes' Discourse on Method,* p. 14.

14   "As soon as I had finished the course of studies": Descartes, *Discourse,* p. 5.

14   "nothing solid could have been built": Descartes, *Discourse,* p. 8.

14   "I did nothing but roam": Descartes, *Discourse,* p. 59.

14   He spent time serving in two armies: A. C. Grayling, in his 2005 biography, speculates that Descartes was working as a spy for the Jesuits, which would explain how he came to be in so many places of military and political significance; Grayling acknowledges, however, that there is no actual evidence for this.

15   "there will remain almost nothing else": Adam and Tannery, *Oeuvres,* vol. 10, p. 158. Quoted in Gaukroger, *Soft Underbelly of Reason,* p. 93.

16   "the father not just of modern philosophy": Schouls, *Descartes and the Possibility of Science,* p. 3.

16   "the dividing line in the history of thought": Roth, *Descartes' Discourse on Method,* p. 3.

18   "The best way of proving the falsity": Adam and Tannery, *Oeuvres,* vol. 9B, p. 18. Quoted in Schouls, *Descartes and the Possibility of Science,* p. 9.

18   "the lords and masters of nature": Descartes, *Discourse,* discourse 6.

19   "The single design to strip one's": Descartes, *Discourse,* p. 49.

19   "that I am in this place": Ibid., p. 119.

20   In this way, Descartes became: A sampling of cogito riffs through the ages: "I think, therefore I spam" (blogger Amitai Givertz). "I blink, therefore I am" (slogan of the Cartesian Reflex Project at the University of Massachusetts/Amherst, which is devoted to the study of involuntary blinking, a subject Descartes wrote about in 1649). "Cogito, ergo Cartesius est" (Saul Steinberg). "I am because my little dog knows me" (Gertrude Stein). "Cogito, ergo spud./I think, therefore I yam" (Internet graffito). "Coito, ergo cum" (poet Gustavo Pérez Firmat). "I stink, therefore I am" (many iterations). And my favorite: "I think, therefore I am. I think" (George Carlin).

20   "Doubt is the beginning": Verbeek, *Descartes and the Dutch,* p. 39.

21   Shortly after, Regius penned a letter: The letter is in Adam and Tannery, *Oeuvres,* vol. 2, p. 305. I am relying on the commentary in Bos, *Correspondence,* pp. 3–9.

22 The chattering classes: Regius, in Bos, *Correspondence*, p. 3.

22 "greatest of all philosophers": Heereboord, in Verbeek, *Descartes and the Dutch*, p. 40. The next quote is from Anton Aemilius, in Bos, p. 18, and the quote about the method is from Regius, in Bos, p. 3.

23 He delved into details: Bos, *Correspondence*, pp. 214–20.

24 "A Sponge to Wash Away": Gaukroger, *Descartes*, p. 359.

24 "learned ignorance": Verbeek, *Descartes and the Dutch*, p. 18.

26 "neither in public nor in private lessons": Quoted in Verbeek, *Descartes and the Dutch*, p. 83.

26 "There are certain newfangled philosophers": Ibid., p. 49.

28 "doubts might be imported": Ibid., p. 39.

31 "But I have found nothing": Letter to Marin Mersenne, in Descartes, *Philosophical Writings*, vol. 3, p. 134.

31 "Now I am dissecting the heads": Letter to Mersenne, in Adam and Tannery, *Oeuvres*, vol. 1, p. 263, quoted in Gaukroger, *Descartes*, p. 228.

31 "prevented by the brevity of life": Descartes, *Discourse*, p. 46.

31 As absurd as Descartes' hopeful ideas: For the discussion of Descartes' hopeful beliefs about medicine, and how he was looked upon for his medical wisdom, I rely mainly on Shapin, "Descartes the Doctor: Rationalism and Its Therapies."

32 Meanwhile, proving that he was truly modern: Adam and Tannery, *Oeuvres*, vol. 1, p. 434.

32 We happen to know that on October 15, 1634: Information on Helena Jans and Francine comes from Baillet, *La vie de Monsieur Des-Cartes*, vol. 2, pp. 89–91; Adam and Tannery, *Oeuvres*, vol. 1, pp. 299, 393–94; Rodis-Lewis, *Descartes*, pp. 137–41; Gaukroger, *Descartes*, pp. 294–95; Van der Ven, "Quelques données."

35 "as a peasant does with his field": Quoted in Masson, *Queen Christina*, p. 144.

35 She was said to treat certain young ladies: Buckley, *Christina*, p. 141.

38 "Perhaps this will pass": Adam and Tannery, *Oeuvres*, vol. 5, p. 430.

39 "I am out of my element": Ibid., p. 467.

39 "that if he had to die": Ibid., p. 478.

40 "oily": Baillet, *La vie de Monsieur Des-Cartes*, vol. 2, p. 420.

40 "the death-rattle, black sputum": Adam and Tannery, *Oeuvres*, vol. 5, pp. 477–78. Translated for the author by Jane Alison.

40 "We are afflicted in this house": The letters quoted in this paragraph are all found in Adam and Tannery, *Oeuvres*, vol. 5, pp. 470–78.

41   "would have lived five hundred years": Baillet, *La vie de Monsieur Des-Cartes*, vol. 2, p. 452. Quoted in Shapin, "Descartes the Doctor," p. 141.

Chapter 2 Banquet of Bones

43   "Death," the philosopher Ludwig Wittgenstein once wrote: The complete passage from the *Tractatus Logico-Philosophicus* goes as follows: "Death is not an event in life: we do not live to experience death. If we take eternity to mean not infinite temporal duration but timelessness, then eternal life belongs to those who live in the present. Our life has no end in just the way in which our visual field has no limits."

46   Where Chanut had been an enthusiastic promoter": My sources on Terlon are Terlon's *Mémoires du Chevalier de Terlon;* the Swedish encyclopedia *Nordisk familjebok*; and Dominique Terlon, family genealogist.

46   Anne attended Mass: My source on Anne of Austria is Kleinman, *Anne of Austria.*

47   Terlon had been in the process: Principal sources on disinterment in Sweden and transportation of the bones from Stockholm to Paris are Baillet, *La vie de Monsieur Des-Cartes*; Adam and Tannery, *Oeuvres*; "Documentation concernant le crâne de Descartes."

48   "nearly the whole Catholic Church of Sweden": Baillet, *La vie de Monsieur Des-Cartes*, vol. 2, p. 598.

49   "without indecence": Ibid., p. 597.

49   Here in Sweden: Lindborg, *Descartes i Uppsala*, p. 339.

52   "the appearance of a bundle of rocks": Adam and Tannery, *Oeuvres*, vol. 12, p. 599.

56   The notes taken by an anonymous lawyer: Clair, *Jacques Rohault*, pp. 51–52.

57   Here Descartes and his followers: I am generalizing here; in fact, there were many variations on this basic Scholastic theme regarding knowledge and perception.

57   Rohault dismantles this logic: This example from Rohault's *System of Natural Philosophy* comes by way of Watson, *Breakdown of Cartesian Metaphysics*, p. 87.

58   A notion had first struck Descartes: Gaukroger, *Descartes*, p. 356.

59   Protestants (some of them, anyway): Lutherans, for one, did not, and

do not, believe either that the host represents the body of Christ or that its substance is replaced by the substance of Christ's body, but rather that, with the act of consecration, the two substances—bread and body—coexist. My account of transubstantiation and Cartesianism relies on Schmaltz, *Radical Cartesianism*; Watson, *Breakdown of Cartesian Metaphysics*; and Armogathe, " 'Hoc Est Corpus Meum' " and *Theologia Cartesiana*.

59 **In Catholic theology:** Precisely what happens during transubstantiation has remained a sticky point for the Catholic Church. In one of its most recent attempts at clarification, a 1981 report on points of reconciliation with the Anglican Church notes, "The word *transubstantiation* is commonly used in the Roman Catholic Church to indicate that God acting in the eucharist effects a change in the inner reality of the elements. The term should be seen as affirming the *fact* of Christ's presence and of the mysterious and radical change which takes place." The document then skirts the problem that gave the Church—and the Cartesians—headaches in the seventeenth century, adding that "in contemporary Roman Catholic theology [transubstantiation] is not understood as explaining *how* the change takes place" (Anglican–Roman Catholic International Commission, *Final Report*, p. 14).

59 **"our Lord Jesus Christ":** Waterworth, *Council of Trent*, 13th sess., ch. 1.

60 **"Numerous stories were known":** Watson, *Breakdown of Cartesian Metaphysics*, p. 160.

62 **"I was so surprised by this":** Gaukroger, *Descartes*, p. 290.

64 **Robert Desgabets:** Details about Desgabets come from Schmaltz, *Radical Cartesianism*, and Watson, *Breakdown of Cartesian Metaphysics*.

64 **Desgabets journeyed to Paris:** Desgabets held that the Aristotelian explanation of the Christian miracle didn't work because it required the first substance, that of the bread, to be eliminated before it could be replaced by the second substance, that of Christ's body. Matter, Desgabets argued, could not be destroyed.

68 **Finally, on an evening in late June:** Details on the funeral of 1667 come from Baillet, *La vie de Monsieur Des-Cartes*, vol. 2, ch. 23.

70 **"illustrious and learned ghost":** Van Damme, "Restaging Descartes."

71 **The archbishop of Paris:** Quotations are from Schmaltz, *Radical Cartesianism*, pp. 29–33.

72 **The end result, however:** Armogathe and Carraud, "L'ouverture des archives de la Congrégation pour la doctrine de la foi."

73   **Most significant for history:** My sources on Malebranche and Arnauld include Lawrence Nolan, "Malebranche's Theory of Ideas and Vision in God," and Steven Nadler, *Arnauld and the Cartesian Philosophy of Ideas.*

Chapter 3 Unholy Relics

77   **"at Cards Dice":** Quoted in Uglow, *Lunar Men,* p. 51.
78   **Voltaire, the godfather of the French Enlightenment:** Voltaire, *Lettres philosophiques,* pp. 2–3.
78   **"was the founder":** Rée, *Descartes,* pp. 30–31.
81   **In the 1720s, Alberto Radicati:** Jacob, *Radical Enlightenment,* pp. 172–76, and Israel, *Radical Enlightenment,* p. 69.
82   **Even among the first generation of Cartesians:** My paragraph on women and female sexuality relies on Israel, *Radical Enlightenment,* ch. 4: "Women, Philosophy, and Sexuality."
83   **"if I vindicate":** Collins, *A Discourse of Free-Thinking,* p. 5.
84   **"the Devil is intirely banish'd":** Ibid., p. 28.
84   **The modern French scholar:** Vovelle, *Piété baroque et déchristianisation en Provence au XVIII siècle.*
84   **"By the *Universe*":** Quoted in Berman, *Atheism in Britain,* p. xii.
85   **"calls God Nature":** Quoted in Israel, *Radical Enlightenment,* p. 4.
85   **"Those who have seen naked spirits":** Ibid., p. 375.
86   **"Immense pains have therefore":** Spinoza, *Theologico-Political Treatise,* preface, in *Chief Works of Benedict de Spinoza.*
87   **"The spirit that animated the reformers":** Quoted in Schouls, *Descartes and the Enlightenment,* p. 73.
89   **"Enlightenment," he declared:** Quoted in Cassirer, *Philosophy of the Enlightenment,* p. 163.
90   **"The Old World imagined":** Commager, *Empire of Reason,* p. xi.
92   **He had a face that could best:** Information on Lenoir and the Museum of French Monuments comes from portions of a doctoral dissertation on Lenoir by Jennifer Carter; correspondence with Carter; Christopher Greene's "Alexandre Lenoir and the Musée des monuments français during the French Revolution"; Lenoir's own *Description historique et chronologique* and *Notice historique;* Louis Courajod's *Alexandre Lenoir,* Guy Cogeval and Gilles Genty's *La logique de l'inaltérable;* and the three-volume *Statistique monumentale de Paris,* compiled by Lenoir's son Albert.

93   Some of the parlements: Quotations and details come from duc de Croy, *Journal inédit du duc de Croy, 1718–1784,* pp. 220–28, and Palmer, *Age of the Democratic Revolution,* pp. 94–96.

99   atheism was proclaimed: A reaction to the extremes of the Cult of Reason sprang up in the form of the Cult of the Supreme Being, which followed a deist course in believing that a divine being, which could be understood by reason rather than faith, oversaw the earth, and France in particular. Meanwhile, at a time when priests were forbidden to practice Mass or give Communion, villagers throughout the country took the sacraments up themselves, with a layman officiating. They tried to get around both the death penalty that was instituted for consecrating a host and Catholic qualms about nonpriests trying to effect the transubstantiation of the host by performing what were called white masses, in which the layman performing the mass did not consecrate the bread and wine but rather instilled in them symbolic meaning. All of which shows, perhaps, that even within the hardened core of radical modernity, the French Revolution itself, the three ways of dealing with the intersection of reason and faith that have played out over and over since the time of Descartes—radical secular, moderate, and determinedly religious—manifested themselves.

102   "he had understood that it must be derived": Quoted in Schouls, *Descartes and the Enlightenment,* p. 67.

104   Lenoir collected feverishly: *Archives du Musée des monuments français,* vol. 2, p. 36.

105   So meticulous was Lenoir: Ibid., pp. 27–37.

106   For Lenoir, Descartes was not only: Lenoir, *Description,* p. 243.

106   "I was a real republican": *Archives du Musée des monuments français,* vol. 1, pp. 16–17.

107   Lenoir later said: Lenoir, *Notice historique,* pp. 22–23.

108   "Your committee of public instruction": Chénier, *Rapport fait à la Convention nationale, au nom du Comité d'instruction publique.*

109   "We have thought that a nation": Quoted in Bonnet, *Naissance du Panthéon,* p. 315.

112   " 'Tis not contrary to reason": Hume, *Treatise of Human Nature,* p. 167.

113   "Man himself must make": Kant, *Religion within the Limits of Reason Alone.*

115   "Sepulchral lamps hang": Lenoir, *Description,* pp. 93–94; here quoted in Greene, "Alexandre Lenoir," p. 213.

116   "The French cherish this famous revolution": Lenoir, *Description*, p. v.

117   "In that calm and peaceful garden": Ibid., p. 17; here quoted in Greene, "Alexandre Lenoir," p. 214.

118   "No. 507. Sarcophagus, in hard stone": Lenoir, *Description*, p. 243.

119   "the remarkable question": The quotation from Chénier and Mercier come from Chénier, *Rapport fait par Marie-Joseph Chénier*, as reprinted in the *Gazette nationale, ou le Moniteur universel* of May 14, 1796, and the chapter "Panthéonisé" that Mercier subsequently wrote in *Le nouveau Paris*.

119   "the 10th of Prairial": In their zeal to transform and modernize every aspect of life the revolutionaries streamlined the calendar and renamed the months. Each revolutionary month had thirty days—divided into three ten-day weeks—and the months were named for seasonal changes. Prairial—which ran from mid-May to mid-June—meant essentially "the month the prairies flower." Of course, it was also imperative to drop the Gregorian dating system, which counted years from the birth of Christ. The new system started time over, with Year One reckoned from the official beginning of the French Republic, in 1792.

120   "those who have invitations": Quoted in Shaw, "Time of Place."

124   He tried to resell it: Lenoir, *Description*, p. 113.

Chapter 4 The Misplaced Head

130   By the end of the 1600s: Coleman, *Georges Cuvier*, p. 18.

131   The greatest of these chemists: My information on Berzelius comes from Jorpes, *Jac. Berzelius*, and from documents in the collection of the Natural History Museum in Paris.

132   "it is impossible to describe the bliss": Jorpes, *Jac. Berzelius*, p. 42.

132   As Maurice Crosland: My account of the Academy of Sciences draws on Crosland, as well as on the academy's own history.

133   Crosland argues that it was: It was also true that the French Revolution nudged the academy out of existence for a time; because it had had royal backing, it was seen as a reactionary institution. It returned as the National Institute, then, in 1816, resumed as the Academy of Sciences.

133   When Franz Mesmer came to Paris: My account of Mesmer is

based on Donaldson, "Mesmer's 1780 Proposal for a Controlled Trial," and on e-mail correspondence with Donaldson.

134   He was awed by Paris: Jorpes, *Jac. Berzelius,* p. 82.

138   He did, however, discuss these observations: I infer that discussion was informal because I haven't been able to locate reference to an official report on the topic in the minutes of the academy during this time, whereas there are later official reports on the remains of Descartes.

142   "has realized all those combinations": Cuvier, *Leçons d'anatomie comparée;* quoted in Coleman, *Georges Cuvier,* pp. 171–72.

144   On April 30, 1821: Details of this meeting come from Académie des Sciences, *Procès-verbaux des séances.*

151   The two astronomers: Delambre had actually crossed paths with Descartes' bones at this earlier time; while the revolutionary leaders were debating moving Descartes' remains to the Pantheon, he was performing his metric measurements from the highest point in Paris—the cupola of the Pantheon.

152   "I have not told the public": Quoted in Alder, *Measure of All Things,* p. 6.

154   Within the first 128 days: Vass, "Beyond the Grave," p. 191.

156   In the 1860s and 1870s: What follows relies on the Liljewalch collection at the Lund University library, Sweden; Verneau, "Les restes de Descartes"; and Ahlström et al., "Cartesius' Kranium." I infer that Liljewalch began to chart Descartes' skull in the 1860s and 1870s from the dates found in his notes: the latest dates for which he recorded notes are 1869 and 1872.

157   He became the object of affection: My miniportrait of Nordenflycht comes from Stålmarck, *Hedvig Charlotta Nordenflycht,* and from the Web site of Uppsala University.

157   another Swedish devotee: My account of Sparrman's career relies on Beaglehole, *Life of Captain James Cook,* and Sparrman, *Voyage to the Cape of Good Hope.*

160   The story sounded quite fascinating: Quoted in Ahlström et al., "Cartesius' Kranium," p. 35.

162   Besides reporting anatomical particularities: What follows comes from Ahlström et al., "Cartesius' Kranium."

165   "the Academy of Sciences received last Monday": Berzelius, *Berzelius brev. I,* pp. 76–84.

Chapter 5 Cranial Capacity

167   **Franz Joseph Gall was nothing if not consistent:** My account of Gall and the advent of phrenology comes from Colbert, *Measure of Perfection*; Lanteri-Laura, *Histoire de la phrénologie*; Staum, *Labeling People*; Young, *Mind, Brain, and Adaptation*; Zola-Morgan, "Localization of Brain Function"; and Ackerknecht, "Contributions of Gall."

168   **"This doctrine concerning the head":** Quoted in Zola-Morgan, "Localization of Brain Function," p. 364.

173   **"I frequently quote Descartes":** Ibid., p. 375.

175   **"Your *faculty*":** Quoted in Young, *Mind, Brain, and Adaptation*, p. 71.

175   **"Each succeeding age":** Ibid., pp. 71–72.

175   **"Descartes goes off to die":** Ibid., p. 72.

176   **"Flourens' advocacy of physiological":** Ibid., pp. 73–74.

177   **"The reason I believe this":** Adam and Tannery, *Oeuvres*, vol. 3, pp. 19–20.

177   **Descartes had scarcely aired:** My account of mind-body dualism is based in part on Lokhorst, "Descartes and the Pineal Gland."

177   **"It does not seem to me":** Adam and Tannery, *Oeuvres*, vol. 3, p. 693; quoted in Lokhorst, "Descartes and the Pineal Gland."

178   **"For many philosophers":** Nagel, *View from Nowhere*, pp. 7–8.

179   **"Although there is a connection":** Ibid., p. 4.

180   **"we cannot expect a physiological":** Académie des Sciences, *Procès-verbaux des séances*, April 25, 1808.

188   **In *Lessons in Comparative Anatomy*:** Cuvier, *Leçons d'anatomie comparée*, lesson 8, p. 7.

188   **"The Negro race":** Cuvier, *Le règne animal*, p. 95.

190   **Wagner identified the root:** Hagner, "Skulls, Brains, and Memorial Culture," p. 210.

192   **"This preference is without doubt":** *Bulletins de la Société d'anthropologie de Paris*, 1861, p. 139.

193   **"In general, the brain is larger":** Quoted in Pearce, "Louis Pierre Gratiolet," p. 263.

195   **"One can thus affirm":** *Bulletins de la Société d'anthropologie de Paris*, 1861, p. 428.

195   **"If it is permitted":** The account of Gratiolet's adventure with Cuvier's hat is found in the *Bulletins de la Société d'anthropologie de Paris*,

1861, p. 428. I have also relied on the retelling of it in Gould, *Panda's Thumb*, and in Schiller, *Paul Broca*.

197 "This is one of the most beautiful types": *Bulletins de la Société d'anthropologie de Paris*, 1861, p. 70.

198 "Monsieur Gratiolet . . . has said": Ibid., p. 71.

198 "Our colleague Monsieur Gratiolet": Ibid., pp. 164–65.

199 "the circumstances of the death of Descartes": Ibid., pp. 224–25.

200 "However, when I expressed this opinion": Ibid., pp. 238–39.

202 "Race is a social concept": "Do Races Differ? Not Really, Genes Show," *New York Times*, August 22, 2000.

202 "In new studies and reviews": http://psychology.uwo.ca/faculty/rushton_res.htm.

203 "inherently gloomy": *Times Online*, October 17, 2007; http://www.timesonline.co.uk/tol/news/uk/article2677098.ece.

Chapter 6 Habeas Corpus

209 The members received a report: Académie des Sciences, *Comptes rendus*, September 23, 1912.

210 "Great turmoil": Verneau, "Les restes de Descartes."

211 "This communication arouses": *Journal des débats politiques et littéraires*, September 25, 1912.

211 "One knew that the great philosopher": *Gazette de France*, September 23, 1912.

211 "Every hypothesis was allowed": Verneau, "Les restes de Descartes."

212 "with all the respect due": Perrier, "Sur le crâne dit 'de Descartes.'"

212 "They talked about it": Cabanès, "Les tribulations posthumes de Descartes."

213 The panel was found: Information on Paul Richer and his encounter with the skull comes from Richer, "Sur l'identification du crâne supposé de Descartes," *Physiologie artistique*, and *L'art et la médicine*; "Le crâne de Descartes," *Le Soir*, January 21, 1913; "Sketch Identifies Skull of Descartes," *New York Times*, January 26, 1913; and Comar, *Mémoires de mon crâne*.

215 "Sketch Identifies Skull of Descartes": *New York Times*, January 26, 1913.

215   "The Skull of Descartes Is Authentic": *Le Figaro*, January 21, 1913.

217   "Descartes was internationally mourned": Slive, *Frans Hals*, vol. 1, p. 164.

221   "The proposal to transfer the ashes": "Pantheon Awaits Descartes Ashes When Discovered."

222   "Where are the remains of Descartes?": *Le Temps*, December 17, 1927.

222   Once this was realized, the matter was dropped: To be precise, there was one recent attempt to authenticate the remains. In 2005, Bernard Cartier, a retired French medical doctor and historian of French science who in the course of doing research on Paul Richer became infected with a similar Cartesian doubt, had the notion to verify Richer's methods using the most modern standard: to dig up the remains at St.-Germain-des-Prés and to perform DNA tests on them, as well as on the skull at the Museé de l'homme. Cartier received official authorization from the permanent secretary of the French National Academy of Medicine "to study regarding the remains of Descartes, at the museum and at the church of St.-Germain des Prés, the feasibility of an investigation into their authenticity." He contacted the appropriate authorities and received replies from the office of the mayor of Paris and the Prefecture of Police describing the translation to the church as, for these officials, adequate proof of authenticity, noting that the tomb was sealed with stone and cement, suggesting the difficulties of jackhammering into an ancient abbey, and urging Cartier, politely, decisively, to let it rest.

222   There are something like thirty-one wars: http://www.human securitybrief.info/.

226   "the circumstances of that time": *Archives du Musée des monuments français*, vol. 2, p. 298.

226   "I pray you, Monsieur": Adam and Tannery, *Oeuvres*, vol. 12, pp. 624–25.

227   "I hasten to respond to the letter": Ibid., pp. 625–27.

230   "to the southern end of the nave": Ibid., p. 602.

Chapter 7 A Modern Face

233 He describes his normal work: Harashima, "The Concealing Face, the Nameless Face."

235 "a symbol of homo sapiens": Interview with the author.

236 The teeth—in particular the wisdom teeth: Information about the skull and the exhibit comes from the exhibit catalog and from interviews and e-mail exchanges with Hisao Baba.

237 "Descartes laid the foundations": Watson, *Cogito, Ergo Sum,* p. 3.

238 One bright winter day: My information about Hirsi Ali comes largely from her autobiography, *Infidel,* and from my own interview with her.

241 "The age which venerated reason": Cassirer, *Philosophy of the Enlightenment,* p. xii.

242 "(1) man is not natively depraved": Becker, *Heavenly City of the Eighteenth-Century Philosophers,* pp. 102–03.

243 "Western culture is also fundamentalist": Schlaffer, "Holiday from the Enlightenment."

244 "We distrust anything": Hitchens, *God Is Not Great,* pp. 5–8.

245 "we are at war with Islam": Van Bakel, "The Trouble Is the West."

Epilogue

251 But a few years ago: Van de Ven, "Quelques données nouvelles sur Helena Jans."

252 "A troop of theologians": Adam and Tannery, *Oeuvres,* vol. 5, pp. 15–16.

Bibliography

Académie des Sciences. *Procès-verbaux des séances de l'Académie tenues depuis la fondation de l'Institut jusqu'au mois d'août 1835*. 10 vols. Hendaye: Imprimérie de l'Observatoire d'Abbadia, 1910–22.

———. *Comptes rendus hebdomadaires des séances de l'Académie des sciences*. Paris: Gauthier-Villars, 1912.

"Académie des sciences: Le crâne de Descartes est authentique." *Le Figaro,* January 21, 1913.

Ackerknecht, E. H. "Contributions of Gall and the Phrenologists to Knowledge of Brain Function." *The History and Philosophy of Knowledge of the Brain and Its Functions*. Amsterdam: Israel, 1957.

Adam, Charles, and Tannery, Paul, eds. *Oeuvres de Descartes*. 12 vols. Paris: Librairie Philosophique J. Vrin, 1974.

Ahlström, Carl Gustaf, Per Ekström, and Ove Persson. "Cartesius' Kranium." *Sydsvenska medicinhistoriska sällskapets årsskrift 1983*.

Åkerman, Susanna. *Queen Christina of Sweden and Her Circle: The Transformation of a Seventeenth-Century Philosophical Libertine*. Leiden: Brill, 1991.

Alder, Ken. *The Measure of All Things: The Seven-Year Odyssey That Transformed the World*. London: Little, Brown, 2002.

Andersson, Ingvar. *A History of Sweden.* London: Weidenfeld and Nicolson, 1956.

Anglican–Roman Catholic International Commission. *The Final Report.* Oxford: Bocardo and Church Army Press, 1981.

Archives du Musée des monuments français. 3 vols. Paris: E. Plon, Nourrit et Cie, 1883–97.

Armogathe, Jean-Robert. *Theologia Cartesiana: L'explication physique de l'Eucharistie chez Descartes et dom Desgabets.* The Hague: Martinus Nijhoff, 1977.

———. "La sainteté janséniste." *Histoire des saints et de la sainteté chrétienne.* Eds. Francesco Chiovaro et al. vol. 9. Paris: Hachette, 1987.

———. " 'Hoc Est Corpus Meum': Le débat autour de l'explication Cartésienne de la transubstantiation eucharistique." *Travaux du Laboratoire européen pour l'étude de la filiation.* Ed. Pierre Legendre. Brussels: Émile Van Balberghe Libraire et Yves Gevaert Éditeur, 1998.

——— and Vincent Carraud. "La première condamnation des *Oeuvres* de Descartes, d'après des documents inédits aux archives du Saint-Office." *Nouvelles de la République des Lettres* 2 (2001).

———. "L'ouverture des archives de la Congrégation pour la doctrine de la foi." *Communio* 30 (January–February 2005).

Aston, Nigel. *Religion and Revolution in France, 1780–1804.* London: Macmillan, 2000.

Baillet, Adrien. *La vie de Monsieur Des-Cartes.* 2 vols. Paris: Daniel Horthemels, 1691.

Baker, Keith Michael. *Condorcet: From Natural Philosophy to Social Mathematics.* Chicago: University of Chicago Press, 1975.

Balz, Albert G. A. "Clerselier (1614–1684) and Rohault (1620–1675)." *The Philosophical Review* 39 (September 1930).

Beaglehole, J. C. *The Life of Captain James Cook.* Palo Alto: Stanford University Press, 1992.

Becker, Carl. *The Heavenly City of the Eighteenth-Century Philosophers.* New Haven: Yale University Press, 1932.

Berman, David, ed. *Atheism in Britain.* Vol. 1: *An Answer to Mr. Clark's Third Defence of His Letter to Mr. Dodwell,* by Anthony Collins, and *A Discourse of Free-Thinking,* by Anthony Collins. Bristol: Thoemmes Press, 1996.

Bernard, Leon. *The Emerging City: Paris in the Age of Louis XIV.* Durham: Duke University Press, 1970.

Berzelius, Jac. *Berzelius brev. I, Brewäxling mellan Berzelius och C.L. Berthollet, 1810–1822.* Stockholm, 1912.

Blagdon, Francis William. *Paris as it was and as it is, or a Sketch of the French capital illustrative of the effects of the Revolution: with respect to sciences, literature, arts, religion, education, manners and amusements, comprising also a correct account of the most remarkable national establishments and public buildings, in a series of letters written by an English traveller during the years 1801–2 to a friend in London.* London: C. A. Baldwin, 1803.

Boddington, A., A. N. Garland, and R. C. Janaway. *Death, Decay and Reconstruction: Approaches to Archaeology and Forensic Science.* Manchester: Manchester University Press, 1987.

Boileau-Despréaux, Nicolas. "Arrêt burlesque." *Oeuvres complètes de Boileau.* 4 vols. Paris: Garnier, 1873.

Bonnet, Jean-Claude. *Naissance du Panthéon: Essai sur le culte des grands hommes.* Paris: Fayard, 1998.

Bos, Erik-Jan. *The Correspondence between Descartes and Henricus Regius.* Utrecht: Zeno, 2002.

———. "Descartes' *Lettre apologétique aux magistrats d'Utrecht:* New Facts and Materials." *Journal of the History of Philosophy* 37 (July 1999).

Buckley, Michael J. *Denying and Disclosing God: The Ambiguous Progress of Modern Atheism.* New Haven: Yale University Press, 2004.

Buckley, Veronica. *Christina, Queen of Sweden: The Restless Life of a European Eccentric.* London: Fourth Estate, 2004.

Cabanès, Dr. "Les tribulations posthumes de Descartes," *Gazette medicale de Paris,* November 6, 1912.

Carter, Jennifer. "Recreating the Poetic Imaginary: Alexandre Lenoir and the Musée des Monuments français." Doctoral dissertation, McGill University, 2007.

Cassirer, Ernst. *The Philosophy of the Enlightenment.* Princeton: Princeton University Press, 1951.

Chénier, Marie-Joseph de. *Rapport fait à la Convention nationale au nom du Comité d'instruction publique, par Marie-Joseph Chénier, suivi du décret rendu à la séance du 2 octobre 1793 (sur le transport au Panthéon du corps de Descartes).* Paris: Imprimérie Nationale, 1793.

———. *Rapport fait par Marie-Joseph Chénier sur la translation des cendres de René Descartes au Panthéon. Séance du 18 floréal l'an IV.* Paris: Imprimérie nationale, 1796.

"Chronique scientifique." *La Gazette de France,* September 23, 1912.

Clair, P., ed. *Jacques Rohault, 1618–1672: Bio-bibliographie, avec l'édition critique des entretiens sur la philosophie.* Paris: CNRS, 1978.

Clark, Gregory. *A Farewell to Alms: A Brief Economic History of the World.* Princeton: Princeton University Press, 2007.

Cogeval, Guy, and Gilles Genty. *La logique de l'inaltérable: Histoire du Musée des monuments français.* Paris: Imprimérie Nationale, 1993.

Colbert, Charles. *A Measure of Perfection: Phrenology and the Fine Arts in America.* Chapel Hill: University of North Carolina Press, 1997.

Coleman, William. *Georges Cuvier, Zoologist: A Study in the History of Evolution Theory.* Cambridge: Harvard University Press, 1964.

Collins, Anthony. *A Discourse of Free-Thinking.* London, 1713.

Comar, Philippe. *Mémoires de mon crâne—René Des-Cartes.* Paris: Gallimard, 1997.

Commager, Henry Steele. *The Empire of Reason.* New York: Doubleday, 1977.

Condorcet, Marquis de (Jean-Antoin-Nicolas de Caritat). *Esquisse d'un tableau: Historique des progrès de l'esprit humain*. Paris: Agasse, 1798.

Courajod, Louis. *Alexandre Lenoir, son journal et le Musée des monuments français*. 3 vols. Paris: H. Champion, 1878–87.

Cousin, Victor. *Madame de Sablé*. Paris: Didier, 1869.

"Le crâne de Descartes." *Le Soir*, January 21, 1913.

"Le crâne de Descartes." *Le Temps*, October 2, 1912.

Crosland, Maurice. *Science under Control: The French Academy of Sciences, 1795–1914*. Cambridge: Cambridge University Press, 1992.

Croy, duc de. *Journal inédit du duc de Croy, 1718–1784*. Vol. 2. Paris: Flammarion, 1907.

Cuvier, Georges. *Leçons d'anatomie comparée*. Paris: Baudoin, 1805.

———. *Le règne animal distribué d'après son organisation, pour servir de base à l'histoire naturelle des animaux et d'introduction à l'anatomie comparée*. Brussels: Culture et civilisation, 1969.

Damasio, Antonio R. *Descartes' Error: Emotion, Reason, and the Human Brain*. New York: G. P. Putnam's Sons, 1994.

Descartes, René. *Discours de la Methode pour bien conduire sa raison, & chercher la verité dans les sciences. Plus La Dioptrique. Les Meteores. Et La Geometrie. Qui sont des essais de cete Methode*. Leyden: Ian Maire, 1637.

———. *Discourse on Method*. Trans. Laurence Lafleur. New York: Macmillan, 1960.

———. *The Philosophical Writings of Descartes*. 3 vols. Ed. J. Cottingham, R. Stoothoff, and D. Murdoch. Cambridge: Cambridge University Press, 1991.

Descartes et les Pays-Bas. Amsterdam: Maison Descartes, 1985.

"Documentation concernant le crâne de Descartes." Dossier compiled by Philippe Mennecier, Muséum national d'histoire naturelle, Musée de l'homme, Laboratoire d'anthropologie biologique, October 1996.

Donaldson, I. M. L. "Mesmer's 1780 Proposal for a Controlled Trial to Test His Method of Treatment Using 'Animal Magnetism.' " *Journal of the Royal Society of Medicine* 98 (2005).

Droge, Jan. *Kasteel Endegeest: Een geschiedenis van het huis, de tuin en de bewoners.* Leiden: Matrijs, 1993.

Dulaure, J. A. *Nouvelle description des curiosités de Paris.* Paris: Lejay, 1785.

Etlin, Richard A. *The Architecture of Death: The Transformation of the Cemetery in Eighteenth-Century Paris.* Cambridge: MIT Press, 1984.

French, Roger, and Andrew Wear. *The Medical Revolution of the Seventeenth Century.* Cambridge: Cambridge University Press, 1989.

Gaukroger, Stephen. *Descartes: An Intellectual Biography.* Oxford: Clarendon, 1995.

———, John Schuster, and John Sutton, eds. *Descartes' Natural Philosophy.* London: Routledge, 2000.

Gaukroger, Stephen, ed. *The Soft Underbelly of Reason: The Passions in the Seventeenth Century.* London: Routledge, 1998.

Gordon, Daniel, ed. *Postmodernism and the Enlightenment: New Perspectives in Eighteenth-Century French Intellectual History.* New York: Routledge, 2001.

Gould, Stephen Jay. *The Panda's Thumb: More Reflections in Natural History.* New York: Norton, 1992.

Goupille, André. *Haya, La Haye en Touraine, La Haye Descartes, Descartes: Des origines à nos jours.* Tours: Chavanne, 1980.

Grayling, A. C. *Descartes: The Life of René Descartes and Its Place in His Times.* London: Free Press, 2005.

Greene, Christopher. "Alexandre Lenoir and the Musée des monuments français during the French Revolution." *French Historical Studies* 12 (1981).

Grell, Ole Peter, and Andrew Cunningham, eds. *Religio Medici: Medicine and Religion in Seventeenth-Century England.* Aldershot: Scolar Press, 1996.

Hagner, Michael. "Skulls, Brains, and Memorial Culture: On Cerebral Biographies of Scientists in the Nineteenth Century." *Science in Context* 16 (2003).

Harashima, Hiroshi. "The Concealing Face, the Nameless Face: Has the Media Really Been Evolving? A Perspective of Facial Studies." *Natureinterface*, no. 4, http://www.natureinterface.com/e/ni04/P016-021/.

Higonnet, Patrice. *Paris: Capital of the World.* Cambridge: Harvard University Press, 2002.

Hillairet, Jacques. *Dictionnaire historique des rues de Paris.* 2 vols. Paris: Éditions de Minuit, 1964.

Hirsi Ali, Ayaan. *Infidel.* New York: Free Press, 2007.

Hume, David. *A Treatise of Human Nature.* Oxford: Oxford University Press, 2000.

Israel, Jonathan I. *Radical Enlightenment: Philosophy and the Making of Modernity, 1650–1750.* Oxford: Oxford University Press, 2001.

Jacob, Margaret C. *The Radical Enlightenment: Pantheists, Freemasons, and Republicans.* London: George Allen and Unwin, 1981.

Jonas, Raymond. *France and the Cult of the Sacred Heart: An Epic Tale for Modern Times.* Berkeley: University of California Press, 2000.

Jorpes, J. Erik. *Jac. Berzelius: His Life and Work.* Berkeley: University of California Press, 1970.

Kant, Immanuel. *Religion within the Limits of Reason Alone.* Trans. Theodore M. Greene and Hoyt H. Hudson. Chicago: Open Court, 1934.

Kleinman, Ruth. *Anne of Austria, Queen of France.* Columbus: Ohio State University, 1985.

Lanteri-Laura, Georges. *Histoire de la phrénologie: L'homme et son cerveau selon F. J. Gall.* Paris: Presses Universitaires de France, 1970.

"Large Skulls." *New York Times,* August 10, 1879.

Lemoine, Bertrand. *Les Halles de Paris.* Paris: L'Equerre, 1980.

Lenoir, Albert. *Statistique monumentale de Paris.* 3 vols. Paris: Imprimérie Impériale, 1867.

Lenoir, Alexandre. *Notice historique des monumens des arts, réunis au Dépôt national, rue des Petits-Augustins.* Paris: Cussac, 1796, 1797.

————. *Description historique et chronologique des monumens de sculpture réunis au Musée des monumens français*. Paris: Laurent Guyot, 1806.

Lindborg, Rolf. *Descartes i Uppsala*. Stockholm: Almqvist & Wiksell, 1965.

Lokhorst, Gert-Jan. "Descartes and the Pineal Gland." *The Stanford Encyclopedia of Philosophy*. Winter 2006. Ed. Edward N. Zalta. http://plato .stanford.edu/archives/win2006/entries/pineal=gland.

Maccioni Ruju, P. Alessandra, and Marco Mostert. *The Life and Times of Guglielmo Libri: Scientist, Patriot, Scholar, Journalist, and Thief*. Hilversum: Verloren, 1995.

Macdonald, Paul S. "Descartes: The Lost Episodes." *Journal of the History of Philosophy* 40, no. 4 (2002).

Masquelet, A. C. "Rembrandt's *Anatomy Lesson of Dr. Nicolaes Tulp* (1632)." *Maîtrise Orthopédique,* www.maitrise-orthop.com.

Masson, Georgina. *Queen Christina*. London: Secker and Warburg, 1968.

McClaughlin, Trevor. "Censorship and Defenders of the Cartesian Faith in Mid-Seventeenth Century France," *Journal of the History of Ideas,* 40:563–81 (1979).

McGahagan, Thomas A. "Cartesianism in the Netherlands, 1639–1676: The New Science and the Calvinist Counter-Reformation." Ph.D. dissertation, University of Pennsylvania, 1976.

Meige, Henry. *Paul Richer et son oeuvre*. Paris: Masson & Co., 1934.

Mercier, Louis-Sébastien. *Éloge de René Descartes*. Paris: Vve Pierres, 1765.

————. *Corps législatif. Conseil des Cinq-Cents. Discours de L.-S. Mercier, prononcé le 18 floréal, sur René Descartes*. Paris: Imprimérie Nationale, 1793.

————. *Le nouveau Paris*. 1799. Paris: Mercure de France, 1994.

Mouy, Paul. *La développement de la physique cartésienne, 1646–1712*. Paris: Librairie Philosophique J. Vrin, 1934.

Nadler, Steven. *Arnauld and the Cartesian Philosophy of Ideas*. Princeton: Princeton University Press, 1989.

Nagel, Thomas. *The View from Nowhere.* New York: Oxford, 1986.

———. "Conceiving the Impossible and the Mind-Body Problem." *Philosophy* 73 (July 1998).

Nolan, Lawrence. "Malebranche's Theory of Ideas and Vision in God." *Stanford Encyclopedia of Philosophy,* Winter 2003.

Nordenfalk, Carl, ed. *Christina, Queen of Sweden: A Personality of European Civilisation.* Stockholm: National Museum, 1966.

Palmer, R. R. *The Age of the Democratic Revolution: A Political History of Europe and America, 1760–1800.* Princeton: Princeton University Press, 1959.

"Pantheon Awaits Descartes Ashes When Discovered." *New York Herald.* European ed. December 3, 1927.

"Pantheon Seeks Descartes' Body." *New York Times,* January 29, 1928.

Parker, M. "False Dichotomies: EBM, Clinical Freedom, and the Art of Medicine." *Medical Humanities* 31 (2005): 25–30.

Pearce, J. M. S. "Louis Pierre Gratiolet (1815–1865): The Cerebral Lobes and Fissures." *European Neurology* 56 (2006).

Pelenski, Jaroslaw, ed. *The American and European Revolutions, 1776–1848.* Iowa City: University of Iowa, 1980.

Perrier, Edmond. "Sur le crâne dit 'de Descartes', qui fait partie des collections du Muséum." *Compte-rendus hebdomadaires des séances de l'Académie des sciences,* September 30, 1912.

Porter, Roy, ed. *Patients and Practitioners: Lay Perceptions of Medicine in Pre-Industrial Society.* Cambridge: Cambridge University Press, 1985.

———, and Mikulas Teich, eds. *The Enlightenment in National Context.* Cambridge: Cambridge University Press, 1981.

"Procès-verbal de la remise à MM. les Commissaires de M. le Préfet de la Seine, des restes de Descartes, Mabillon et Montfaucon, qui étaient déposés dans le Jardin des Petits-Augustins à Paris." *Extrait du Moniteur.* Paris: Agasse, n.d.

"Un Project Vieux de 136 Ans." *La Presse,* November 29, 1927.

Raymond, Jean-François de. *Descartes et Christine de Suède: La reine et le philosophe.* Paris: Bibliothèque Nordique, 1993.

———. *Pierre Chanut, ami de Descartes: Un diplomate philosophe.* Paris: Beauchesne, 1999.

Rée, Jonathan. *Descartes.* London: Allen Lane, 1974.

Rhine, Stanley. *Bone Voyage: A Journey in Forensic Anthropology.* Albuquerque: University of New Mexico Press, 1998.

Richard, Camille. "Le comité de salut public et les fabrications de guerre sous la Terreur." Doctoral dissertation, University of Paris, 1921.

Richer, Paul. *Physiologie artistique de l'homme en mouvement.* Paris: Aulanier, 1896.

———. *L'art et la médecine.* Paris: Gaultier, 1902.

———. "Sur l'identification du crâne supposé de Descartes par sa comparaison avec les portraits du philosophe." *Comptes-rendus hebdomadaires des séances de l'Académie des sciences,* January 20, 1913.

Rodis-Lewis, Geneviève. *Descartes: His Life and Thought.* Ithaca: Cornell University Press, 1995.

Rohault, Jacques. *Oeuvres posthumes de M. Rohault.* The Hague: Henry van Bulderen, 1690.

Roth, Leon. *Descartes' Discourse on Method.* Oxford: Clarendon Press, 1937.

Ruestow, Edward G. *Physics at 17th- and 18th-Century Leiden.* The Hague: Martinus Nijhoff, 1973.

Schiller, Francis. *Paul Broca: Founder of French Anthropology, Explorer of the Brain.* Berkeley: University of California Press, 1979.

Schlaffer, Heinz. "Holiday from the Enlightenment." www.signandsight.com. February 27, 2006.

Schmaltz, Tad M. *Radical Cartesianism: The French Reception of Descartes.* Cambridge: Cambridge University Press, 2002.

Schouls, Peter. *Descartes and the Enlightenment.* Kingston: McGill-Queen's University Press, 1989.

———. *Descartes and the Possibility of Science*. Ithaca: Cornell University Press, 2000.

Scott, Franklin D. *Sweden: The Nation's History*. Minneapolis: University of Minnesota Press, 1977.

Sebba, Gregor. "Some Open Problems in Descartes Research." *Modern Language Notes* 75 (March 1960).

Shapin, Steven, "Descartes the Doctor: Rationalism and Its Therapies." *British Journal for the History of Science* 33 (2000).

Shaw, Matthew. "The Time of Place: Louis-Sébastien Mercier and the Hours of the Day." Paper presented at the Society for the Study of French History Annual Conference, Southampton, England, July 5, 2005.

"Sketch Identifies Skull of Descartes." *New York Times*, January 26, 1913.

Slive, Seymour. *Frans Hals*. 3 vols. Washington: National Gallery of Art, 1989.

Solies, Dirk. "How the Metaphysical Need Outlasted Reductionism: On a Methodological Controversy between Philosophy and the Life Sciences in 19th-Century Germany." Paper presented at the Metanexus Institute's *Continuity + Change: Perspectives on Science and Religion* conference, Philadelphia, June 3–7, 2006.

Sommaire du plaidoyer pour l'abbé, prieur et chanoines réguliers et chapitre de Sainte Geneviève, défendeurs, contre messire Hardouin de Péréfixe, archevêque de Paris, demandeur. Paris, 1667.

Sparrman, Anders. *A Voyage to the Cape of Good Hope, towards the Antarctic Polar Circle, Round the World, and to the Country of the Hottentots and the Caffres from the Year 1772–1776*. Cape Town: Van Riebeeck Society, 1975.

Spinoza, Baruch. *The Chief Works of Benedict de Spinoza*. New York: Dover, 1955.

Stålmarck, Torkel. *Hedvig Charlotta Nordenflycht—Ett Porträtt*. Stockholm: Norstedts, 1997.

Staum, Martin S. *Labeling People: French Scholars on Society, Race, and Empire, 1815–1848*. Montreal: McGill-Queen's University Press, 2003.

Taylor, Quentin. "Descartes's Paradoxical Politics." *Humanitas* 14, no. 2 (2001).

Terlon, Hugue, Chevalier de. *Mémoires du Chevalier de Terlon. Pour rendre compte au Roy, de ses négociations, depuis l'année 1656 jusqu'en 1661.* Paris: Louis Billaine, 1682.

Uglow, Jenny. *The Lunar Men: Five Friends Whose Curiosity Changed the World.* New York: Farrar, Straus and Giroux, 2003.

Van Bakel, Rogier. "The Trouble Is the West." *Reason,* November 2007.

Van Bunge, Wiep. *From Stevin to Spinoza: An Essay on Philosophy in the Seventeenth-Century Dutch Republic.* Leiden: Brill, 2001.

Van Damme, Stéphane. "Restaging Descartes: From the Philosophical Reception to the National Pantheon." http://dossiersgrihl.revues.org/document742.html.

Van de Ven, Jeroen. "Quelques données nouvelles sur Helena Jans." *Bulletin Cartésien* 32 (2001).

Vass, Arpad A. "Beyond the Grave—Understanding Human Decomposition." *Microbiology Today,* November 2001.

Vauciennes, P. Linage de. *Mémoires de ce qui s'est passé en Suède, et aux provinces voisines, depuis l'année 1645 jusques en l'année 1655, tirés des dépêches de Monsieur Chanut par P. Linage de Vauciennes.* Cologne: Pierre du Marteau, 1677.

Venita, Jay. "Pierre Paul Broca." *Archives of Pathology and Laboratory Medicine* 126 (March 2002).

Verbeek, Theo. *Descartes and the Dutch: Early Reactions to Cartesian Philosophy, 1637–1650.* Carbondale: Southern Illinois University Press, 1992.

———. *Une université pas encore corrompue . . . : Descartes et les premières années de l'Université d'Utrecht.* Utrecht: Utrecht University, 1993.

———, Jelle Kingma, and Philippe Noble. *Les néerlandais et Descartes.* Amsterdam: Maison Descartes, 1996.

Verbeek, Theo, Erik-Jan Bos, Jeroen Van de Ven, eds. *The Correspondence of René Descartes: 1643.* Utrecht: Zeno, 2003.

Verneau, R. "Le crâne de Descartes." *L'Anthropologie* 23 (1912): 640–42.

———. "Les restes de Descartes." *Æsculape* 11 (1912): 241–46.

Voltaire. *Lettres philosophiques.* Vol. 2. Paris: Librairie Hachette, 1909.

Vovelle, Michel. *Piété baroque et déchristianisation en Provence au XVIIIe siècle.* Paris: Plon, 1974.

Waterworth, J., trans. *The Canons and Decrees of the Sacred and Œcumenical Council of Trent, Celebrated under the Sovereign Pontiffs Paul III, Julius III, and Pius IV.* London: C. Dolman, 1848.

Watson, Richard A. *The Breakdown of Cartesian Metaphysics.* Indianapolis: Hackett, 1987.

———. *Cogito, Ergo Sum: The Life of René Descartes.* Boston: David R. Godine, 2002.

Wessel, Leonard P. *G. E. Lessing's Theology, a Reinterpretation: A Study in the Problematic Nature of the Enlightenment.* The Hague: Mouton, 1977.

Wilkin, Rebecca M. "Figuring the Dead Descartes: Claude Clerselier's *Homme de René Descartes* (1664)." *Representations* 83 (2003).

Young, Robert M. *Mind, Brain, and Adaptation in the Nineteenth Century.* New York: Oxford University Press, 1990.

Zola-Morgan, Stuart. "Localization of Brain Function: The Legacy of Franz Joseph Gall (1758–1828)." *Annual Review of Neuroscience* 18, (1995): 359–83.

Index

ALSO BY RUSSELL SHORTO

*"Astonishing. . . . A book that will permanently alter the way
we regard our collective past. "*
—The New York Times

THE ISLAND AT THE CENTER OF THE WORLD
*The Epic Story of Dutch Manhattan and the Forgotten
Colony That Shaped America*

When the British wrested New Amsterdam from the
Dutch in 1664, the truth about its thriving, polyglot society
began to disappear into myths about a cartoonish peg-
legged governor and an island purchased for twenty-
four dollars. But the story of the Dutch colony of New
Netherland was merely lost, not destroyed: 12,000 pages
of its records—recently declared a national treasure—are
now being translated. Drawing on this remarkable
archive, Russell Shorto has created a gripping narrative—
a story of global sweep centered on a wilderness called
Manhattan—that transforms our understanding of early
America. The Dutch colony predated the "original" thirteen colonies, yet it seems strikingly familiar. Its capital
was multi-ethnic, and its citizens valued free trade, individual rights, and religious liberty. Their champion was a
progressive young lawyer named Adriaen van der Donck,
who emerges in these pages as a forgotten American patriot and whose clashes with Peter Stuyvesant, the colony's
autocratic director, laid the foundation for New York City
and helped shape American culture.

History/978-1-4000-7867-7

VINTAGE BOOKS
Available at your local bookstore, or visit
www.randomhouse.com